Francesco Bettella

A Linear Regression Approach to Protein Secondary Structure Prediction

Francesco Bettella

A Linear Regression Approach to Protein Secondary Structure Prediction

The History of SPARROW

Südwestdeutscher Verlag für Hochschulschriften

Impressum / Imprint
Bibliografische Information der Deutschen Nationalbibliothek: Die Deutsche Nationalbibliothek verzeichnet diese Publikation in der Deutschen Nationalbibliografie; detaillierte bibliografische Daten sind im Internet über http://dnb.d-nb.de abrufbar.
Alle in diesem Buch genannten Marken und Produktnamen unterliegen warenzeichen-, marken- oder patentrechtlichem Schutz bzw. sind Warenzeichen oder eingetragene Warenzeichen der jeweiligen Inhaber. Die Wiedergabe von Marken, Produktnamen, Gebrauchsnamen, Handelsnamen, Warenbezeichnungen u.s.w. in diesem Werk berechtigt auch ohne besondere Kennzeichnung nicht zu der Annahme, dass solche Namen im Sinne der Warenzeichen- und Markenschutzgesetzgebung als frei zu betrachten wären und daher von jedermann benutzt werden dürften.

Bibliographic information published by the Deutsche Nationalbibliothek: The Deutsche Nationalbibliothek lists this publication in the Deutsche Nationalbibliografie; detailed bibliographic data are available in the Internet at http://dnb.d-nb.de.
Any brand names and product names mentioned in this book are subject to trademark, brand or patent protection and are trademarks or registered trademarks of their respective holders. The use of brand names, product names, common names, trade names, product descriptions etc. even without a particular marking in this works is in no way to be construed to mean that such names may be regarded as unrestricted in respect of trademark and brand protection legislation and could thus be used by anyone.

Coverbild / Cover image: www.ingimage.com

Verlag / Publisher:
Südwestdeutscher Verlag für Hochschulschriften
ist ein Imprint der / is a trademark of
OmniScriptum GmbH & Co. KG
Heinrich-Böcking-Str. 6-8, 66121 Saarbrücken, Deutschland / Germany
Email: info@svh-verlag.de

Herstellung: siehe letzte Seite /
Printed at: see last page
ISBN: 978-3-8381-3819-0

Zugl. / Approved by: Berlin, Freie Universität, Dissertation zur Erlangung des akademischen Grades des Doktors der Naturwissenschaften, 2009

Copyright © 2014 OmniScriptum GmbH & Co. KG
Alle Rechte vorbehalten. / All rights reserved. Saarbrücken 2014

A Linear Regression Approach to Protein Secondary Structure Prediction

Contents

I scope and method   7

1 Introduction   9
- 1.1 Proteins 9
  - 1.1.1 Amino acids 10
  - 1.1.2 Protein structure 10
    - 1.1.2.1 Secondary structure 12
  - 1.1.3 Protein structure prediction 14
    - 1.1.3.1 Secondary structure prediction .. 15
    - 1.1.3.2 State of the art 15
- 1.2 Scope of this project 18

2 The method   19
- 2.1 Overview 19
  - 2.1.1 Prediction instructions 19
    - 2.1.1.1 A comparative statistical learning method 19
    - 2.1.1.2 Defining the problem 20
- 2.2 Data 21
  - 2.2.1 Structural information 21
- 2.3 Program design 22
  - 2.3.1 Learning module 23
    - 2.3.1.1 Learning data set 24
    - 2.3.1.2 Validation data set 24
    - 2.3.1.3 Recognition 24
    - 2.3.1.4 The library of secondary structure patterns ... 25
    - 2.3.1.5 Alphabet reduction scheme 26
    - 2.3.1.6 Relating primary to secondary structure 26
  - 2.3.2 Evaluating module 28
    - 2.3.2.1 Quality measures 29
    - 2.3.2.2 Statistical relevance 32

3 Scoring functions   34
- 3.1 Sequence vector 34
  - 3.1.1 Standard representation 36

	3.1.2	Profile representation	36

	3.2	Single-state likelihood functions	37
		3.2.1 Function optimization procedure	38
		3.2.1.1 Parametrical candidate estimators	38
		3.2.1.2 Least squares	39
		3.2.1.3 Linear equation system	40
	3.3	From single-state to multi-state	41

II development   43

4 Standard sequence-based predictor   45

	4.1	Linear scoring functions	45
		4.1.1 Dual statistics	50
		4.1.2 Filtering the learning data set	54
		4.1.2.1 Positive reinforcement	54
		4.1.2.2 SSE length filters	56
		4.1.3 The secondary structure key-position	57
		4.1.4 The secondary structure motifs unleashed	60
		4.1.5 Focusing on single-choice classifications	64
		4.1.5.1 Structure-dependent optimization schemes	64
		4.1.5.2 Multi-choice classification tests	68
	4.2	Quadratic scoring functions	68
		4.2.1 Reducing overfitting	72
		4.2.2 Clustering amino acids	72
		4.2.3 Extended secondary structure patterns	76
		4.2.3.1 Initialized prediction	77
		4.2.3.2 Gapped secondary structure patterns	80

5 Profile-based predictor   81

	5.1	Preliminary tests: a quality leap	81
		5.1.1 Choosing a reduction scheme	82
	5.2	Super scoring functions	83
		5.2.1 The score neighbourhood	84
		5.2.2 Quadratic super scoring functions	85
		5.2.3 Further upgrades	86
		5.2.4 Test results	88
		5.2.4.1 Optimizing the regularization parameters	88
		5.2.4.2 Estimating performance	93
		5.2.4.3 Taking a closer look	94
		5.2.4.4 The missing correlations	97

III applications 101

6 The application software 103
6.1 SPARROW 103
6.1.1 Introducing transmembrane domains 104
6.1.1.1 Two special membrane proteins 105
6.1.2 Prediction confidence 105
6.2 SPARROW versus others 108
6.2.1 ASTRAL40 generations: a neutral testing ground 108
6.2.2 Non-aligned sequences 110
6.2.3 A 3D picture of secondary structure prediction 113
6.2.4 SPARROW's output examples 120

7 Conclusions and outlook 131

IV appendices 135

A Single-state prediction accuracy 137

B SPARROW prediction data 142

Notation

It might be useful to keep the following notations in mind. Some of them may be, for the sake of clarity, formally introduced in the body of the report.

- $\mathbb{N}$ — set of all natural numbers
- $\mathbb{Z}$ — set of all integer numbers
- $\mathbb{R}$ — set of all real numbers
- $M(n, m, \mathbb{R})$ — set of all real-valued $(n \times m)$-matrices
- $M(n, \mathbb{R})$ — set of all real-valued $(n \times n)$-matrices
- $\mathbb{1}_n$ — $n \times n$ identity matrix
- λ — scalar quantity
- $\mathbf{w}$ — vector in the sequence vector space
- $\boldsymbol{\varepsilon}$ — vector in the sample space
- δ_{ij} — Kronecker symbol
- $\langle x \rangle$ — mean of x, typically over a number samples
- $\langle x \rangle_{\pm}$ — weighted mean of x over $+$ and $-$ samples
- $S, \mathcal{S}$ — typically a set
- $|\mathcal{S}|$ — cardinality of set $\mathcal{S}$
- $\{a, b, c\}$ — set containing elements a, b, c
- (a, b) — open interval from a to b
- $[a, b]$ — closed interval from a to b
- $\mathcal{F}(A, B)$ — set of functions defined in A with values in B

Part I
scope and method

Chapter 1

Introduction

Biology is the science of life.

But what is life? Despite its widespread use, there is no general agreement as to what exactly the notion "life" stands for [1, 2]. Giving credit to some appealing arguments originated in the science of complexity [3–6] a living entity could in general be thought of as one presenting a complex, "interesting" behaviour, as opposed, for instance, to chaotic, fluid-like, or rigidly ordered, crystal-like ones[1].

The most interesting behaviours on planet Earth in this sense, as far as the human scientific community can tell, is observed in a class of self-organized [7, 8] chemical systems characterized by a number of common features, the most striking of which is that of being based on a double helical molecule called deoxyribonucleic acid, or, in short, DNA.

1.1 Proteins

The DNA appears to be a sort of library that contains the instructions for the self-organization and self-sustainment of the chemical systems called "living" organisms, the instructions to construct their other basic components: *proteins*.

Proteins are linear chains of covalently bonded molecules called *amino acids*. Their sequences are encoded in DNA segments called *genes*.

Proteins participate in almost all activities that take place within an organism and perform a huge variety of functions. Some of them are *enzymes* that catalyze biochemical reactions, and are vital to metabolism. Others have structural or mechanical functions, such as the proteins of the cytoskeleton, which form a system of scaffolds to maintain a cell's shape. Proteins are also important in processes of the so-called immune response, in cell adhesion, cell signalling, and in the cell cycle.

The sequence of amino acids building up the chain is believed to be solely accountable [9] for the spatial arrangement (fold) of the protein and ultimately for the special role the protein is going to play.

---

[1]The two extremes could be regarded respectively as *entropy* or *energy* driven systems.

1.1.1 Amino acids

An amino acid is a molecule containing both amine and carboxyl functional groups. In biochemistry, however, and for the purposes of this project too, what really go under the name of amino acids are only the 20 standard natural amino acids (see figure 1.1). With the exception of proline, these all adhere to the same template, including an α-carbon to which the amine and the carboxyl groups and a variable side-chain are bonded. What drives the folding process and thus leads to the final three-dimensional structure of the protein are the different physicochemical properties of the side-chains.

The amino acids in a protein are linked by peptide bonds formed in a dehydration reaction. For this reason, proteins are often called *peptides* (or *polypeptides* if they are particularly long), though the former name is preferred when referring to the complete biological molecule in its final stable conformation or *native state*. Once linked in the protein chain, an amino acid is called a *residue* and the linked series of carbon, nitrogen and oxygen atoms are known as *main chain* or protein *backbone*. The end of the protein with a free carboxyl group is commonly known as the C-terminus, whereas the end with a free amino group is known as the N-terminus.

1.1.2 Protein structure

Most proteins fold into unique three-dimensional structures determined by their *primary* structure, that is, by the sequence of amino acids actually composing them. Assembled together in the native three-dimensional protein structure, the amino acids enlisted in the primary structure organize themselves in regularly recurrent local structural motifs mostly stabilized by means of hydrogen bonds. The most common examples of such structural motifs are *alpha-helices* [10] and *beta-strands* [11]. The local arrangements of a polypeptide chain are collectively called *secondary* structure, while the way in which the polypeptide chain (eventually locally organized in secondary structure domains) finally folds in the three-dimensional space is called *tertiary* structure. The latter is generally stabilized by non-local interactions, most commonly by the formation of a hydrophobic core, but also through hydrogen bonds, disulphide bonds and salt bridges. Finally, in many cases, two or more polypeptide chains, called in this context protein *subunits*, can form larger complexes, which then constitute what is commonly regarded as the protein's *quaternary* structure. The definitions of protein primary, secondary, tertiary and quaternary structure were first introduced by Kaj Ulrik Linderstrøm-Lang in 1952 [12].

It is widespread belief that in their transition to the native state, the amino acid sequences take on roughly the same route and proceed through the same intermediate states (see figure 1.2). The folding process seems to involve the es-

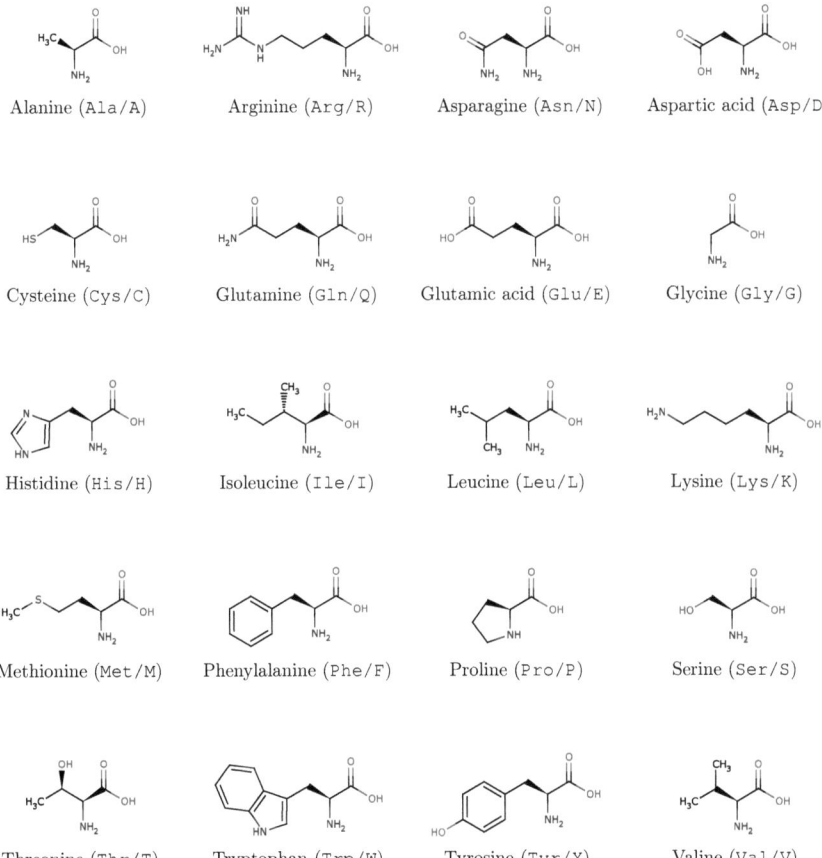

Figure 1.1: The 20 standard natural amino acids in their skeletal representation. In parentheses are respectively their three-letter and one-letter codes. As can be seen, proline deviates from the scheme the other amino acids adhere to, in that its N-end nitrogen is involved in an unusual ring with the side-chain. This, incidentally, makes proline technically an *imino* acid rather than an amino acid.

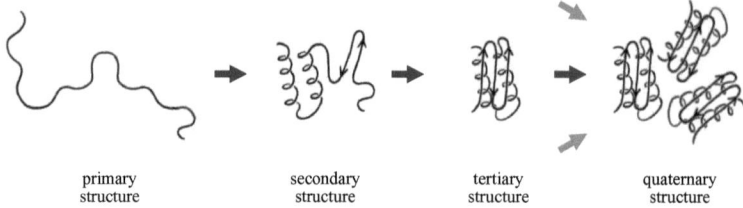

| primary structure | secondary structure | tertiary structure | quaternary structure |

Figure 1.2: Schematic view of the possible path of a polypeptide chain to its native state. The chain arranges itself locally in secondary and supersecondary structures which subsequently fold into the tertiary conformation. Several folded subunits can then come together in the quaternary conformation.

tablishment of regular secondary and supersecondary[2] structural features first [13–17], particularly α-helices and β-sheets, and afterwards of tertiary ones. Formation of quaternary structure usually involves the assembly of subunits that have already folded.

Despite the regularity with which they fold, proteins are not rigid molecules. On the contrary, while performing their biological function they very often shift between several different structures. The various tertiary or quaternary structures they adopt are usually referred to as *conformations*, and the transitions between these are called conformational changes. Such changes are often induced by the binding of a substrate molecule to an especially "sensitive" region of the protein called *active site*, or by interaction with other proteins. In addition to conformational changes, all proteins in solution undergo constant variations in structure due to thermal vibration and association with other molecules. In the context of a protein's relatively dynamic life the secondary structure domains appear to enjoy instead a rather stable existence and while they do often play a central role in the protein's activity itself, they see their structural integrity for the most part unaffected.

1.1.2.1 Secondary structure

The secondary structure of a protein is defined by patterns of hydrogen bonds between backbone amide and carboxyl groups. The hydrogen bonding patterns are, however, correlated with other structural features as well and this has given rise to alternative definitions of secondary structure. The most famous is the geometrical definition based on the dihedral angles ϕ and ψ formed by the backbone α-carbons, that is, the angles between the planes formed by two consecutive

---

[2]Supersecondary structures involve the association of several secondary structures motifs in a particular geometric arrangement. Examples of supersecondary structure are hairpins, corners and even Rossmann folds.

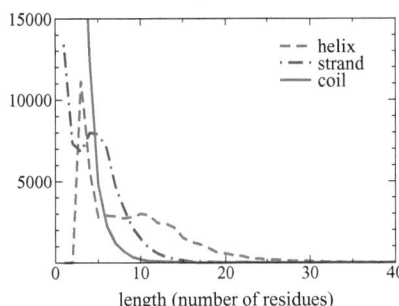

Figure 1.3: Distribution of the standard secondary structure elements as a function of their length. This distribution was obtained from the release 1.71 of the AsTRAL40 data set [19]. The secondary structures were determined using DSSP (see section 2.2.1) and a loose reduction scheme (see section 2.3.1.5) to project the secondary structure types resolved by DSSP to the three standard ones (helix, strand and coil).

peptide bonds. Since the secondary structure formation induces specific arrangements of the backbone, embodied by specific values of the dihedral angles, a segment of residues with such dihedral angles is sometimes called a "helix" or a "strand" regardless of whether it exhibits the correct hydrogen bonding pattern or not. The correlation between secondary structure and dihedral angles is commonly summarized in the Ramachandran plot [18]. Many other definitions have been proposed, often applying concepts from the differential geometry of curves, such as curvature and torsion.

The secondary structure content of a protein can be estimated spectroscopically. A common method is the one based on far-ultraviolet circular dichroism. Less common but still widely used is the method based on infrared spectroscopy, which detects differences in the bond oscillations of amide groups due to hydrogen bonding. In addition, secondary structure contents may as well be accurately estimated using the chemical shifts of an unassigned NMR spectrum.

The most common secondary structure types are α-helices and β-strands. Other types of helices, such as the 3-turn helix and 5-turn helix (also called π-helix), are known to have energetically favourable hydrogen-bonding patterns as well but are far less frequently observed in natural proteins. Short β-strands bridging residues in distant portions of a protein chain are often quite appropriately called β-bridges and assigned a class of their own. Some structural features like tight hydrogen-bonded turns and loose, flexible loops often link the more "regular" secondary structure elements (SSEs) as is the case, for instance, in α-helix bundles (see for example figure 6.6a) and β-sheets (see for example figure 6.8b), series of parallel or antiparallel β-strands. Finally, the motif called "random coil" is not a true secondary structure type as such, but rather the class of conformations that indicates the absence of any other regular secondary structure. Figure 1.3 shows the distribution of helices, strands and coils as a function of their length in number of residues.

Amino acids vary considerably in their ability to form the various secondary structure elements. Proline and glycine residues are sometimes known as "helix breakers" because they disrupt the regularity of the α-helical backbone conformation. Both have unusual conformational abilities and are often found in turns. Amino acids that prefer to adopt helical conformations in proteins include alanine, leucine, methionine, glutamine, glutamic acid and lysine; by contrast, the large aromatic residues (tryptophan, tyrosine and phenylalanine) and the C$_\beta$-branched amino acids (isoleucine, valine, and threonine) tend to adopt β-strand conformations. At any rate, these preferences are not strong enough to constitute alone a reliable criterion for predicting secondary structure. In relatively recent times, several different methods have been developed to do just this.

1.1.3 Protein structure prediction

Since the structure of a protein plays such a central role for its function within the living organism, it is subject of great scientific interest. Several experimental techniques have been developed in order to investigate it, the most popular of which are X-ray crystallography and NMR spectroscopy. These techniques have given great insight into many mechanisms involving proteins, and have thus helped to shed light on countless processes essential for life. They are however relatively expensive and time consuming. As a consequence, especially considering the massive amount of protein data derived from modern large scale DNA sequencing, the need is increasingly felt for *theoretical* prediction methods providing a means of generating plausible structures for those proteins the structure of which is yet unknown. Medicine and biotechnologies also provide important fields of application for protein structure prediction. This can be a very precious aid in fact in designing drugs and other functional or structural molecules.

Predicting the structure of proteins is a very difficult task. The number of atoms involved in such molecules is most of the times too large to allow a thorough "ab initio" electrostatic computation, let alone a quantum mechanical one. A direct simulation of protein folding in atomic detail via methods like molecular dynamics on the other hand is to rule out too, due to its very high computational cost. For these reasons, most structure prediction methods actually rely on more or less simplified representations of proteins in which amino acids are typically treated as structureless units rather than in atomic detail.

Given the huge number of possible amino acid combinations ($N = 20^L$ for a sequence comprising L amino acids), an exhaustive inspection of the entire sequence space is not feasible. So, for example, in the frame of ab initio techniques the *likelihood* of a certain fold can be established by means of "energy" functions [20, 21] and contact maps [22] that summarize in a way the interaction between the amino acid side-chains. Of a somewhat different nature are *homology modelling* and *protein threading* techniques. The former [23] is based on the reasonable assumption that two homologous proteins, that is proteins with "similar" sequences, will share very similar structures. Because a protein's fold is evolutionarily more conserved than its amino acid sequence, a target sequence can be moulded with reasonable accuracy on a very distantly related template, provided that the relationship between target and template can be discerned through sequence alignment. The latter [24] consists in scanning the amino acid sequence with unknown structure against the database of solved structures. A scoring function not dissimilar from the energy functions of ab initio methods is then used to assess the sequence's compatibility with each candidate structure (also called *decoy*), thus yielding suitable three-dimensional models.

A considerable aid is provided to all these structure prediction methods if a prediction of the secondary structure [25–27] of the protein is available beforehand. Reliable secondary structure information may in fact be employed to build up safe starting cores in fold simulation programs [28] and especially, as effective structural constraints for protein threading [29, 30] and homology modelling [31–33] searches.

1.1.3.1 Secondary structure prediction

Early secondary structure prediction methods were based on the mentioned alpha and beta propensities of single amino acids [34–36] or exploited certain physicochemical properties of these [37]. Such methods reached an accuracy ratio of about 0.65 in predicting which of three states, helix, strand or coil (see Q_3 index definition in section 2.3.2.1) a residue adopts. The main drawback of these methods was that they were taking into account no contextual information [38, 39].

In the years the field has grown considerably and so has the secondary structure prediction accuracy. Several different methods have been developed, varying in more respects, from the information actually taken into account to the adopted techniques. The latter include information theory [40–45], neural networks [46–62], homology and nearest neighbour searches [63–71] eventually coupled to a physicochemical analysis [72, 73], hidden Markov models [68, 74–77], stochastic tree grammars [78], support vector machines [79–84] not to mention multiple linear regression analysis (a technique not dissimilar from the one to be presented in this report) [85], genetic algorithms [86] and various combinations of such techniques [87, 88]. Some have also worked out ways to merge together a number of existing tools in the frame of so-called *consensus* methods [89–94], often succeeding in increasing the overall prediction quality.

The strongest performance boost was given to secondary structure prediction programs by the introduction of multiple sequence alignments and profiles to replace the pure sequence as input material for the prediction program.

1.1.3.2 State of the art

Unfortunately, it is not so straightforward to compare the prediction qualities achieved by various methods because they were often tested in different conditions. A solution to this problem might have been represented by the EVA (Evaluation of Automatic protein structure prediction) project [95–97]. Drawing on pre-release data from the Protein Data Bank (PDB) [98], EVA provided in fact a universal, continuous and statistically significant benchmark platform for all secondary structure prediction servers that took part in it. According to EVA's statistics, using EVA's own "standard of truth" (see section 2.3.1.5), the average accuracy ratio in predicting which of the three secondary structure states, helix, strand or coil, a residue adopts (see section 2.3.2.1) appeared to be somewhere below 0.8.

The EVA project seems to be no longer active at the time of writing but still represents a valuable reference for the secondary structure community. The servers officially benchmarked by EVA are listed in table 1.1 together with other prediction programs available in the internet.

Following are some further details about them.

- APSSP2 [99] combines a modified example based learning (EBL) technique with a neural network approach.

- PHD [100], like other prediction tools, is based on a system of neural networks. Its main merit was that of introducing profiles in secondary structure prediction.

- Porter's bidirectional recurrent neural networks [56] make use of long range information to improve the predictor's accuracy.

- PROF [90] by Ouali and King combines GOR-like and neural network methods.

- Prospect [101] is designed to predict the fold of a protein my means of threading techniques. It can input secondary structure information or predict it on its own.

- PSIPRED [51] uses a bilayer neural network and two-stage learning. It was the first to make use of PSIBLAST-generated profiles.

- SABLE2 [58] uses neural networks to combine secondary structure *and* relative solvent accessibility (RSA) predictions.

- SAM [102] is an homology-based predictor. The homologs to a target sequence are found by means of a hidden Markov model which is iteratively constructed in the learning process.

- SSpro4 [55] is the precursor of Porter.

Table 1.1: List of active secondary structure prediction programs as of April 24th, 2009. The last column indicates whether a version of the prediction program was also available for download. In the first part are servers which were tested by EVA; in the second part, those which were no longer or never tested by EVA.

NAME	METHOD	REF.	DOWNLOAD
\multicolumn{4}{c}{servers benchmarked in EVA}			
APSSP2	nearest neighbour and neural network	[99]	no
PHD	neural network	[100]	yes
Porter	neural network	[56]	no
PROF	consensus	[90]	yes
Prospect	threading	[101]	yes
PSIPRED	neural network	[51]	yes
SABLE2	neural network	[58]	yes
SAM	hidden Markov models	[102]	yes
SSpro4	neural network	[55]	yes
YASPIN	neural network and hidden Markov models	[54]	no
\multicolumn{4}{c}{servers *not* benchmarked in EVA}			
BSMPSSP	segmental semi-Markov models	[103]	no
GOR	information theory	[42, 44, 93]	yes
IPSSP	hidden semi-Markov models	[74]	no
JPred	neural network	[104]	no
JUFO	neural network	[105]	no
MUPRED	nearest neighbour and neural network	[88]	no
PMSVM	support vector machine	[82]	no
PREDATOR	homology and physical chemistry	[73]	yes

- YASPIN's architecture is not dissimilar from that of other bilayer neural networks. What makes it different is that the second stage prediction is performed in its case by a hidden Markov model instead of a second neural network [54].

- BSMPSSP [103] uses a generalization of hidden Markov models to treat the secondary structure prediction as a general Bayesian inference problem.

- GOR-IV and GOR-V [42, 44, 93] are the successors of the original GOR (Garnier-Osguthorpe-Robson) method, based on probability parameters derived from empirical studies of known protein structures.

- IPSSP's architecture is very close to that of BSMPSSP; it prides itself, however, of *not* making use of profiles, but of what the authors call *single* amino acid sequences [74].

- JPred [104], the evolution of Jnet, uses juries of neural networks in a sort of *internal* consensus approach.

- JUFO's neural network architecture differs from others in that it accepts also low-resolution non-local tertiary structure information as an additional input [105]. In most application this would come from *de novo* tertiary structure prediction methods like ROSETTA [106].

- MUPRED [88] uses a neural network to combine the strengths of fuzzy k-nearest neighbour [107] prediction and profiles.

- PMSVM [82] adopts a two-stages learning procedure similar to the one adopted by many neural network-based methods but replaces the neural networks themselves with support vector machines.

- PREDATOR [73] generates secondary structure propensities for a target sequence *and* a set of sequences related to it by sequence alignment.

1.2 Scope of this project

The aim of the project reported on here was to develop a new method for predicting a protein's secondary structure on the basis of its primary structure.

The method developed prides itself of originality and a relative straightforwardness and eventually proved to be a competitive alternative to those existing at the time of writing.

Chapter 2
The method

2.1 Overview

The general diagram of a sequence-based secondary structure predictor is depicted in figure 2.1.

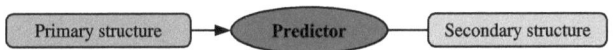

Figure 2.1: Basic flow diagram of a sequence-based secondary structure predictor. The amino acid sequence, or primary structure, is input by the program which returns its secondary structure prediction.

2.1.1 Prediction instructions

In order for a prediction program to have any real predicting potential, it must be instructed on how to make best use of the input it receives. This is done here by constructing a set of *rules* to correlate a protein's primary structure with its secondary structure. This process, accompanied by the quality assessment of the achieved predicting capability, was the main concern of the project and shall be the subject of the rest of this report.

The following subsections present an overview of the type of technique used to instruct the predictor.

2.1.1.1 A comparative statistical learning method

The method developed belongs to the broader class of *comparative* techniques. Techniques of this kind *use available information* to infer a set of more or less manifest *rules* which somehow identify the system of interest. These rules can be either implicitly applied, like in methods based on direct homology searches, or used in a semi-empirical fashion to build *ad hoc* models describing the system of interest. Such models differ from those stemming from *ab initio* techniques in that they are not based on any pre-established theoretical constructs and are generally very specialized. They are often so entangled in the program that generates them, that it is hard, if not impossible, to identify the rules that actually define them. This is the case, for instance, for multilayer artificial neural networks. This is not the case, however, for models like the one illustrated here. As will be shown in detail in section 5.2.4.3, the parameters contained in the predictor described here have a very precise significance and can be directly related to the input it receives, that is, the sequence of amino acids. This is one of the advantages of this predictor compared to those based on other more obscure learning machines.

The "statistical" attribute characterizing the method in the title of the thesis as well as in this section delineates the nature of the rules-inferring procedure. This is based, as the reader shall see, on the statistical analysis of existing data and fits in the theoretical framework defined by Vapnik's statistical learning theory [108].

2.1.1.2 Defining the problem

It is quite widespread practice in the bio-informatics community to represent the secondary structure of a protein as a string of characters with the same length as the primary structure, in terms of number of amino acids, and assign in this way a secondary structure motif to every single residue in the chain.

In the following example, the beta chain of E. coli heat-labile enterotoxin [109], amino acids are represented by their one-letter code and "H" stands for helix, "E", signifying here "extended", stands for strands and "." stands for coil, which is not, as already pointed out, a secondary structure motif in itself but rather means "no structure", or, at least, "none of the above structures".

```
TITELCSEYR  NTQIYTINDK  ILSYTESMAG  KREMVIITFK
SGETFQVEVP  GSQHIDSQKK  AIERMKDTLR  ITYLTETKID    primary
KLCVWNNKTP  NSIAAISMKN                            structure


.HHHHH....  .EEEEEEE.   ..EEEE...   ...EEEEE..
...EEEE...  .......HHH  HHHHHHHHHH  HHHH....EE    secondary
EEEEE.....  EEEEEEEEE.                            structure
```

Given a certain amino acid sequence, the predictor will return a string like the second one. The quality of this can be straightforwardly evaluated by simple direct comparison with the true[1] one (see section 2.3.2 for further details on the criteria used to evaluate the prediction).

---

[1]The definition of the true secondary structure itself is not as precise as it may seem.

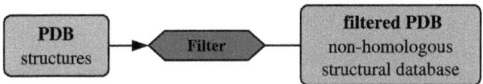

Figure 2.2: Scheme of the PDB filtering process. The filtered database contains representatives of all relevant protein classes.

2.2 Data

In order for the program to be able to deduce any correlation rules between primary and secondary structures, it must have access to the actual secondary structures of the proteins as well as to their amino acid sequences.

The information available to the program is in the form of protein structural data, which can be obtained, for instance, from existing online databases like the Protein Data Bank (PDB) [98].

The PDB contains many protein structures, a fair amount of which share considerable portions of their amino acid sequence with each other. While this might be to a tiny extent due to the natural occurrence likelihood of certain amino acid combinations, determinant factors are in some cases the relative straightforwardness with which some structures can be resolved and often the privileged scientific interest reserved to some protein families. In order to use properly *unbiased* data it is therefore necessary to filter (see figure 2.2) the bulk database, selecting only proteins that are representatives of particular classes.

The data set most extensively employed throughout this project was the release 1.71 of the ASTRAL40 compendium [19, 110, 111], from the SCOP database. This data set contains domain sequences with up to 40% homology.

2.2.1 Structural information

The files in the databases contain the coordinates in space of the atoms (with the possible exception of hydrogen ones) of every amino acid in the protein chain. The protein's secondary structure can be derived from these coordinates using tools like DSSP [112], STRIDE [113] or DEFINE [114].

It is opportune to spend some words about the process of secondary structure determination carried out by such tools. All programs mentioned above make use of PDB coordinates to get information about the geometry and the intramolecular interaction patterns of specific peptide chains. The secondary structure of the peptide is determined on the basis of this information. Plain though the whole procedure may seem, it is not exempt from some shortcomings. The PDB structures, to begin with, are affected by certain degrees of approximation. Another approximation is then introduced by the algorithms, in order to simplify what are

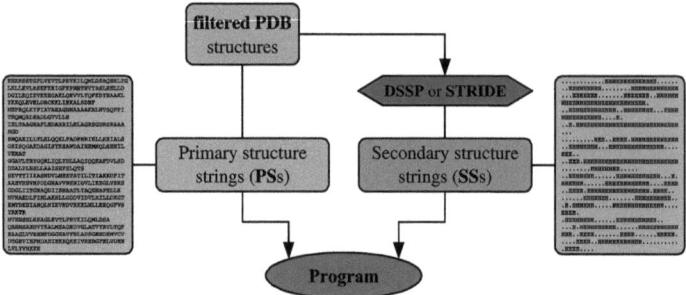

Figure 2.3: Scheme of the data feeding. The primary and secondary structure information is extracted out of the PDB data and passed to the program in the form of strings.

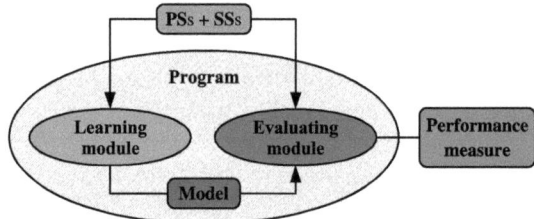

Figure 2.4: Scheme of the program's modular structure. The primary and secondary structure information is passed on to the learning module first. Based on them, the learning module produces a model representing the correlation between primary and secondary structure. This model is then passed on to the evaluating module, which tests its quality.

otherwise too complex interaction networks. DSSP, for example, identifies hydrogen bond patterns using pre-assigned atomic partial charges for the carbonyl oxygen and the amide hydrogen. The consequent non-exactness of the definition of secondary structure intrinsically limits the accuracy attainable by comparative predictors of the kind discussed in this report (see also section 2.3.2.1).

2.3 Program design

Once both primary and secondary structures are available and stored in strings like those seen in the example in section 2.1.1.2, they can be fed to the program as shown in the scheme in figure 2.3.

The program itself is divided into two main modules (figure 2.4):

- **learning module**: constructs a model based on the data;

- **evaluating module**: tests the model prediction performance.

The actual secondary structure predictor tool will be in essence based on the evaluating module save for the quality assessment functionalities themselves.

2.3.1 Learning module

As already pointed out, the learning module is to construct a set of rules correlating the primary and the secondary structures of the protein domains provided.

Unit peptides. The domains in the database contain a few tens to several hundreds amino acids. Handling so heterogeneous data would prove quite impractical. It is more convenient instead, to base the treatment of amino acid sequences not on the whole original protein chains but on unit peptides, extracted out of these (see figure 2.5), containing an equal number of amino acids. A typical unit peptide can contain between 5 and 15 amino acids. In the work reported here, peptides with up to 39 amino acids (see, for instance, section 4.1.5) have occasionally also been considered.

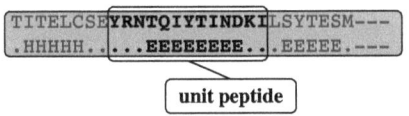

Figure 2.5: Extraction of the primary and secondary structure strings of a unit peptide with 13 amino acids from the original chain. Such peptides are more easily processed by the learning module.

The hypothesis of locality. It is important to point out a significant conceptual repercussion of the use of unit peptides. Aside from its obvious practical advantages, the extraction of a peptide sequence out of its natural protein environment implies ignoring the influence the latter has on the secondary structure of the former. In other words, by basing the analysis on segments of the protein chains rather than on the whole protein chains themselves the secondary structure is assumed to depend only *locally* on the primary structure. Though this doubtless is in many cases a reasonable approximation, there are some situations in which it might definitely prove too rough [115, 116]. The β-sheets for example, formed by rows of parallel or antiparallel β-strands, often involve pairs of residues that are quite far apart in the sequence, but come very close to each other as the protein folds and constitute crucial stability factors for the tertiary structure as well as for the secondary structure itself. It is plausible that neglecting distant correlations such as these might prevent a predictor from identifying a structure that owes in part its very existence to them.

2.3.1.1 Learning data set

Ahead of the actual learning phase a stack of unit peptides like that of figure 2.5, effectively constituting the predictor's knowledge, is selected from the whole data set and the corresponding primary and secondary structure substrings are stored. These are then used to investigate the correlation existing between them.

Supporting data sets. In some cases the learning data set can be further divided into different learning subsets, which are then used to optimize different features of the model. In the case of neural networks, for instance, a portion of the learning data set is left out of the training, but is used to prevent overfitting. As soon as the network's prediction performance on this portion gets worse the training is interrupted. In this work, a fraction ($\sim$10%) of the learning data set, called *supporting* data set, was used to optimize the statistical reweighting factors (see section 4.1.5.1).

2.3.1.2 Validation data set

Since the database of proteins is not unlimited it is imaginable that an algorithm may be devised, which is capable of learning all the information contained in it by heart. For instance, a function that simply maps each position in the primary structure string to the motif at the corresponding position in the secondary structure string would achieve one hundred percent prediction accuracy without even being very resource-demanding. Such a function would however be totally helpless if presented with an amino acid sequence that is *not* included in the database it was "instructed" with; in fact it would not work at all. This is expressed by saying that the "predictor" based on that function has no *generalization* capability and is therefore of no use.

In order to have a more realistic measure of the quality of the model it is advisable to leave out of the learning procedure a portion of the initial database of proteins for cross-validation (figure 2.6). Proteins belonging to this portion are in other words *not* taken into account when establishing the correlation rules and can be used to test the model's actual generalization capability.

2.3.1.3 Recognition

The set of protein domains or unit peptides that were used in the learning procedure can still be tested on to check how well the program has actually learned them. A slightly better performance is to be expected, in general, in the recognition of the learned data set than in the true prediction. A predictor that achieves a very good accuracy in recognition at the expenses of the true prediction is said to be *overfitted* to the learned data set.

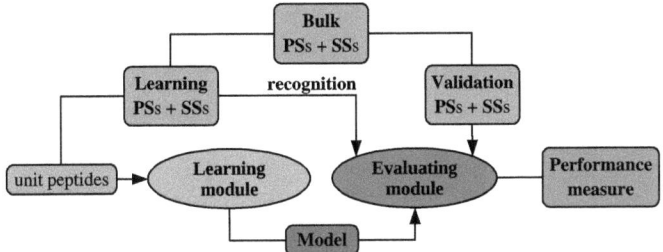

Figure 2.6: Scheme of the data set splittings. The bulk database is divided into two parts, one of which is used for learning, while the other one is used for validation. The "recognition" test, carried out on the former part, tells how well the learning module has learned the data it was given. A test carried out on the latter part gives a measure of the predictor's generalization capability.

2.3.1.4 The library of secondary structure patterns

Before any correlation rules between primary and secondary structure can be established, it is necessary to specify *which* secondary structure patterns the sequence is to be correlated with.

In theory, all the information included in the secondary structure string (or more realistically, in the substring corresponding to the unit peptide currently under examination) provided by the secondary structure definition program can be utilized. In practice though, in order to limit the complexity of the problem, it is advisable to use a few of the secondary structure motif characters at most. Once the number of characters constituting a secondary structure pattern is established, a catalogue of secondary structure patterns can be compiled. This is best illustrated with the help of some examples.

The most typical scenario is the one in which the secondary structure patterns involve a single amino acid position so that the library simply coincides with the secondary structure alphabet $\mathcal{M}$, that is, in general[2], with the alphabet of motifs the secondary structure definition program is capable of sorting out.

Assuming to be using, for instance, DSSP, which is actually the case for most of the results that will be reported, the library of different secondary structures, $\mathcal{P}_1^{(8)}$, would consist of the eight states [112]: α-helix ("H"), isolated β-bridge ("B"), extended strand that participates in β-ladder ("E"), 3-turn-helix ("G"), 5-turn-helix ("I"), hydrogen bonded turn ("T"), bend ("S") and coil (represented in DSSP by a white space).

It is also feasible to consider *extended* patterns (see section 4.2.3), that is,

---

[2]It is possible, as will be shown in section 4.1.4, to further extend the alphabet provided by the secondary structure definition program DSSP [112].

patterns involving more than one residue position. In this case the secondary structure library can grow considerably. If, for example, an alphabet consisting of three basic states (helix "H", strand "E" and coil ".") were to be used and the patterns comprised *two* amino acid positions, the library would be

$$\mathcal{P}_2^{(3)} = \left\{ \begin{array}{lll} \text{HH}, & \text{HE}, & \text{H.}, \\ \text{EH}, & \text{EE}, & \text{E.}, \\ \text{.H}, & \text{.E}, & \text{..} \end{array} \right\},$$

where alongside the *persistence* patterns, "HH", "EE" and "..", also the *transition* patterns, "HE", "H.", and "EH", "E.", ".H" and ".E", make their appearance. Similarly, were the secondary structure involving *three* amino acid positions instead of two, the library would result in

$$\mathcal{P}_3^{(3)} = \left\{ \begin{array}{llllllll} \text{HHH}, & \text{HHE}, & \text{HH.}, & \text{HEH}, & \text{HEE}, & \text{HE.}, & \text{H.H}, & \text{H.E}, & \text{H..}, \\ \text{EHH}, & \text{EHE}, & \text{EH.}, & \text{EEH}, & \text{EEE}, & \text{EE.}, & \text{E.H}, & \text{E.E}, & \text{E..}, \\ \text{.HH}, & \text{.HE}, & \text{.H.}, & \text{.EH}, & \text{.EE}, & \text{.E.}, & \text{..H}, & \text{..E}, & \text{...} \end{array} \right\}.$$

2.3.1.5 Alphabet reduction scheme

The secondary structure definition programs characterize secondary structure based on several (eight in the case of DSSP) different motifs. If a smaller number of secondary structure types is needed, a suitable reduction scheme is necessary to convert the raw secondary structure information. In the secondary structure prediction community, for example, mainly two different reduction schemes are employed to shrink the eight DSSP states down to the three basic ones. The *strict* reduction scheme assigns "H" states to helix and "E" states to extended (or strand) only, leaving everything else to the coil meta-class. Using this scheme the populations of the three basic secondary structures in the release 1.71 of ASTRAL40 amount to 32.7%, 21.1% and 46.2% respectively. The *loose* reduction scheme assigns instead all helix-like states, i.e. "G", "I" and "H", to helix and all strand-like states, i.e. "B" and "E" to strand, partially balancing in this way the secondary structure populations. For the release 1.71 of ASTRAL40 these amount now to 36.4%, 22.2% and 41.4% respectively. Incidentally, the loose reduction scheme also corresponds to EVA's "standard of truth" [96].

2.3.1.6 Relating primary to secondary structure

Given the library of secondary structure patterns the next question arises: how can the amino acid sequence be actually related to these patterns?

Though this clearly is a *multi-choice* classification problem, the approach chosen was to start by looking at all patterns separately and reduce it to the sum of as many so-called *single-choice* classification problems as there are patterns in the library (see figure 2.7). In a single-choice classification problem all the

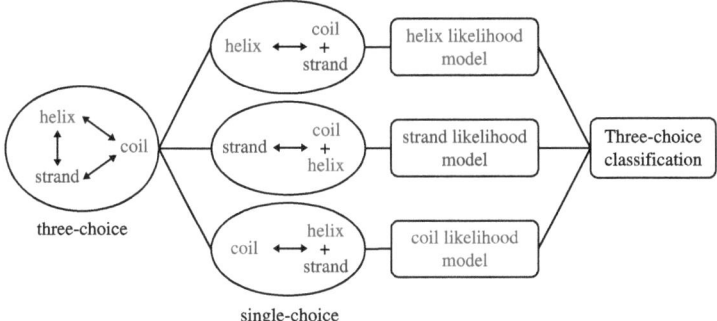

Figure 2.7: Scheme of the separation of the three-choices classification problem into three single-choice classification problems. As the reader will see in the next sections, each single-choice problem gives rise to a likelihood model for the secondary structure of interest. Comparing all likelihood models obtained this way allows to finally address the three-choices problem.

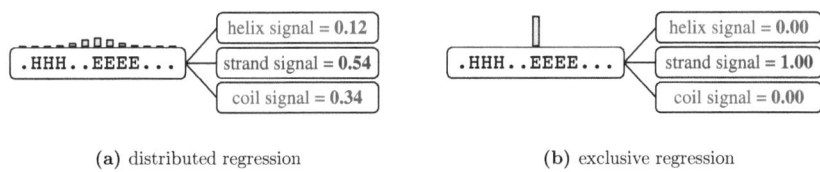

(a) distributed regression (b) exclusive regression

Figure 2.8: Visual representation of the secondary structure signal extraction from a polypeptide segment of length 13 in the case of (a) distributed and (b) exclusive regression. The yellow bars represent different signal intensities. In (a) both strand and coil get fairly strong signals, while in (b) only strand gets a non-zero signal.

patterns that are not the one of interest are grouped into a single class and rules are established, in turn, to sort the one of interest out[3]. Each single-choice classification problem is addressed (see chapter 3) taking inspiration from the procedure[4] outlined by R. A. Fisher in 1936 [118]. It turns out, as will be explained in more detail in section 3, that each set of rules won through Fisher's procedure provides a means of measuring the *likelihood* that some amino acid sequence be associated to the corresponding secondary structure pattern. Once all needed sets of rules are available, a simple direct comparison of the likelihoods attainable from them allows to address the multi-choice classification problem.

---

[3]Strictly speaking it is a double choice, one choice being the secondary structure of interest, the other being all others. I thought however that the "single" attribute better stresses the fact that only *one* secondary structure type is handled at each time.

[4]Incidentally, the same procedure was successfully applied in this group [117] to the problem of the major histocompatibility complex (MHC) binding affinity.

For the sake of establishing the correlation rules two different ways of proceeding were worked out, involving two slightly different regression [85] approaches.

Distributed regression approach. In this case (see figure 2.8a) "signals" are extracted for *all* secondary structure patterns in the library from each peptide sequence in the learning data set. This can be done, for instance, by looking at the secondary structure pattern at different positions along the unit peptide and ponder each contribution. This way, sequence windows containing more occurrences of a same pattern will give, as a rule, a stronger signal than those containing a little bit of everything. These, in turn, will give possibly weak non-zero signals for all patterns present.

Exclusive regression approach. In this case a single secondary structure pattern is associated to each peptide sequence in the learning data set (see figure 2.8b). This is designated by means of a coherent assignment criterion. The criterion used in this work, was to extract the secondary structure pattern at one position along the sequence, elected to be the *key*-position, and assign this pattern to the data sample. This procedure can also be regarded as an extreme case of distributed regression in which one pattern gets all the "signal" while the others get nothing.

The distributed approach was only briefly investigated and promptly abandoned in favour of the exclusive approach which, at least in the context in which the two were compared, lead to slightly better performance.

2.3.2 Evaluating module

The learned correlation rules are put to test here.

A data set of amino acid sequences, in the form of strings like those seen precedently, is input to the program, which performs a prediction based on them. The way this is done differs slightly depending on the type of data set the evaluating module is actually fed with.

Unit-wise evaluation. The data set is composed of unit peptides. This could be the case, for example, if the interest is in ascertaining the model's ability of recognizing the assigned secondary structure pattern. In this case the evaluation procedure consists in applying the correlation rules to every single peptide and compare the predicted secondary structure pattern to the one used (see section 2.3.1.6) during the learning procedure.

Chain-wise evaluation. The data set consists of entire protein domains. Since the correlation rules necessarily involve unit peptides, the procedure effectively consists in isolating subsequent portions of the sequence with the appropriate length (see figure 2.9). These are then fed as unit peptides to the model to get its "suggestions" regarding the secondary structure to be assigned to the amino acids of interest. How these suggestions are finally put together to build the complete protein

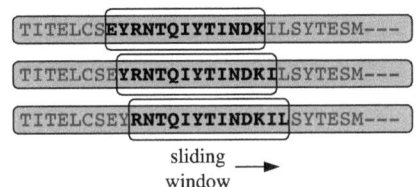

Figure 2.9: Chain-wise evaluation and prediction. A sliding window subsequently isolates the unit peptides to be fed to the evaluating module. The latter gives a suggestion for the secondary structure of the amino acids of interest by applying the learned unit-based rules.

secondary structure string depends very much on the model employed, in the first instance on the size of the secondary structure patterns. Further details on these procedures will therefore be discussed within the chapters dedicated to the predictor's development.

2.3.2.1 Quality measures

Once secondary structure guesses are available for all residues of interest, an overall measure of their quality can be obtained by comparing them with the motifs ascribed by the secondary structure definition program. The most common quality evaluation measures are those based on the three-states quality indices Q_3 and Sov$_3$ (segment overlap [119]), the three states being helix, strand and coil. In this work, only Q_3, reckoned to be more general, was used. A multi-class correlation coefficient (R_m) [120] that more accurately reflects the subtleties of multi-choice classification problems was also employed. Partial accuracy indicators for individual secondary structure motifs also exist, which tell how good the predictor is in detecting a single motif in the context of a multi choice classification problem: here motif-specificity and motif-sensitivity were used. A useful tool to summarize a multi-choice prediction scenario, especially to express the rather elaborate multi-class correlation coefficient, is the so-called *confusion* matrix. Each entry $C_{\sigma\rho}$ in this matrix contains the number of motifs of class ρ predicted to be of class σ, where $\sigma, \rho \in \mathcal{M}$, the standard secondary structure alphabet. For conciseness, the total number of occurrences of motif σ will be called N_σ, while N will stand for the total number of residues for which a prediction is actually performed. Figure 2.10 contains the definitions of the quality indices mentioned above, as functions of these quantities.

It is important to remember that, as has been said before, the problem of secondary structure prediction necessarily requires certain approximations (cf. sections 2.2.1 and 2.3.1). Though the quality measures defined in figure 2.10 do allow for "perfection", this might in reality be unachievable. It has been indeed

- $Q_m = \dfrac{\sum_\sigma C_{\sigma\sigma}}{\sum_{\sigma\rho} C_{\sigma\rho}} = \dfrac{\sum_\sigma C_{\sigma\sigma}}{N}$;

- $\sigma\text{-Spec}_m = \dfrac{C_{\sigma\sigma}}{\sum_\rho C_{\sigma\rho}}$ (σ-specificity);

- $\sigma\text{-Sens}_m = \dfrac{C_{\sigma\sigma}}{\sum_\rho C_{\rho\sigma}} = \dfrac{C_{\sigma\sigma}}{N_\sigma}$ (σ-sensitivity);

- $R_m = \dfrac{\sum_{\sigma\rho\mu}(C_{\sigma\sigma}C_{\rho\mu} - C_{\sigma\rho}C_{\mu\sigma})}{\sqrt{\sum_\sigma\left(\sum_\rho C_{\sigma\rho}\sum_{\mu}\sum_{\nu\neq\sigma} C_{\nu\mu}\right)}\sqrt{\sum_\sigma\left(\sum_\rho C_{\rho\sigma}\sum_{\mu}\sum_{\nu\neq\sigma} C_{\mu\nu}\right)}}$.

Figure 2.10: Definition of multi-choice prediction quality indices. In the formulas, $C_{\sigma\rho}$ are the entries of the confusion matrix (see text), N is the total number of data samples, N_σ is the number of occurrences of class σ and $m = |\mathcal{M}|$ is the number of different classes. See [120] for more details on the correlation coefficient R_m

- **specificity** :
$$\text{Spec} = \frac{\text{TP}}{\text{TP} + \text{FP}}$$

- **sensitivity** :
$$\text{Sens} = \frac{\text{TP}}{\text{TP} + \text{FN}}$$

- **overall accuracy** :
$$Q = \frac{\text{TP} + \text{TN}}{\text{TP} + \text{TN} + \text{FP} + \text{FN}}$$

- **Matthews correlation coefficient (MCC)** :
$$M = \frac{\text{TP} \cdot \text{TN} - \text{FP} \cdot \text{FN}}{\sqrt{(\text{TP} + \text{FN}) \cdot (\text{TP} + \text{FP}) \cdot (\text{TN} + \text{FP}) \cdot (\text{TN} + \text{FN})}}$$

Figure 2.11: Single-choice prediction quality indices definitions. In all the formulas TP indicates the number of unit sequences correctly classified as positive, TN the number of unit sequences correctly classified as negative, FP the number of unit sequences wrongly classified as positive and FN the number of unit sequences wrongly classified as negative.

estimated [121] that the accessible Q_3 prediction accuracy could be intrinsically limited to about 0.88.

Pattern-wise quality control. The various learning procedures aimed at finding the best correlation between the sequence and a single secondary structure class provide an additional frame in which the reliability of the learned rules can be preliminarily surveyed.

In the single-choice classification problem, the unit peptides are divided in two classes which can be called *positive* class and *negative* class. The peptides which have been assigned the current secondary structure pattern of interest belong to the positive class, the others belong to the negative class. Provided some *threshold* likelihood has been previously established, it is possible to infer the program's classification of each unit peptide by comparing the computed likelihood for that peptide to present the pattern of interest, to the established threshold likelihood. The numbers of correctly or incorrectly assigned data samples can then be used to derive additional single-choice prediction quality indices. Figure 2.11 summarizes the ones used in this work.

Table 2.1: ASTRAL40 – release 1.71 subsets. The subsets statistics were computed using both the strict and loose DSSP reduction scheme (see section 2.3.1.5): subsets 1 to 10 contained 704 domains each, while the transmembrane domains amounted to 183. The quantities N_{helix}, N_{strand} and N_{coil} are the number of residues in the corresponding state, N is the total number of residues.

set	DSSP strict			DSSP loose			N
	N_{helix}	N_{strand}	N_{coil}	N_{helix}	N_{strand}	N_{coil}	
set 1	40622	26109	57498	45238	27387	51604	124229
set 2	42238	26316	59410	47057	27772	53135	127964
set 3	41032	26254	57463	45646	27576	51527	124749
set 4	38323	26478	57200	42800	27849	51352	122001
set 5	42439	25622	58649	47196	27070	52444	126710
set 6	39744	26006	57229	44332	27441	51206	122979
set 7	37890	25440	53868	41980	26595	48623	117198
set 8	42998	26954	58469	47770	28269	52382	128421
set 9	40815	26188	58618	45521	27537	52563	125621
set 10	43260	27964	59575	48333	29322	53144	130799
membrane	11591	8571	16210	12937	8996	14439	36372
total	420952	271902	594189	468810	285814	532419	1287043

2.3.2.2 Statistical relevance

Whenever the complexity of the prediction scenario allowed it or the scenario itself was reckoned to be particularly relevant, a thorough statistical analysis of the results was performed. In order to do this, the same procedure was repeated several times with randomly assigned data sets. In this work, two main sets of subdivisions were employed, called respectively $\mathcal{R}_{50/50}$ and $\mathcal{R}_{90/10}$ subdivisions.

The $\mathcal{R}_{50/50}$ set consisted of ten subdivisions obtained by dividing the ASTRAL40 data set into two subsets of roughly equal size (~3600 protein domains), one of which to be used to carry out the functions' optimization procedure, the other to be left out for cross-validation.

In order to generate the $\mathcal{R}_{90/10}$ set instead, the ASTRAL40 data set was first deprived of all transmembrane protein domains. The remaining domains were then divided in ten subsets, nine of which to be used in turn to optimize the scoring functions, the remaining one to be left out for cross-validation. Tests carried out on the subset containing the transmembrane domains will be reported on in section 6.1.1. Table 2.1 contains more detailed information about the ten (+1) subsets.

Calling $\mathcal{R}$ the set of the generated data partitions into learning and validation subsets, the results were averaged out, according to

$$\bar{\mathcal{Q}} = \frac{\sum_{s=1}^{|\mathcal{R}|} \mathcal{Q}^{(s)} \cdot N_s}{\sum_{s=1}^{|\mathcal{R}|} N_s}, \qquad (2.3.1)$$

where $\mathcal{Q}$ stands for any one of the quality indices listed in figures 2.10 and 2.11, $|\mathcal{R}|$ is the cardinality of $\mathcal{R}$, i.e. the number of random partitions, and $\mathcal{Q}^{(s)}$ is the quality index achieved on the N_s residues contained in sample s (or the number of unit peptides in that sample in case of a unit-wise evaluation). If not specified otherwise the deviations shown are computed from the weighted variance

$$\epsilon_Q^2 = \frac{\sum_{s=1}^{|\mathcal{R}|} \left(\mathcal{Q}^{(s)} - \bar{\mathcal{Q}}\right)^2 \cdot N_s}{\sum_{s=1}^{|\mathcal{R}|} N_s}. \quad (2.3.2)$$

Notice that equations (2.3.1) and (2.3.2) reduce to the familiar averages and standard deviations if the random data set assignments are done in such a way as to preserve the total number of residues.

Chapter 3
Scoring functions

As anticipated in the introduction, at the core of the secondary structure predictor developed is a function which, given an amino acid sequence, returns a measure of the likelihood that the amino acids within this sequence adopt one *specific* secondary structure pattern. Such likelihood will later on be referred to also as score, hence the name "scoring function".

The central issue to be addressed in this chapter is how to give shape to such functions, in such a way that they perform their task in the best possible way. The method developed in this project is straightforward. It consists in *analytically* determining the function (of a certain type[1]) that best *fits* the available data.

In order to be able to treat the problem quantitatively, some mathematical notions and tools, like, for instance, the sequence vector space and a space of functions defined on it, must be introduced. The tools will be the same independently from which approach will be pursued, let it be distributed or exclusive regression (see section 2.3.1.6 of the introduction), but the process for designating the scoring functions slightly differs in the two cases. In the following, is briefly explained how.

3.1 Sequence vector

The primary structure can be processed by a computer program in different ways. Often it can be taken as a string of identifiers like those listed in figure 1.1. This is the case, for example, for methods based on sequence alignment searches. Sometimes though, it has to be translated into other better suited mathematical entities.

The predictor to be developed here will be applied on portions of protein sequences, the unit peptides (see section 2.3.1), containing L amino acids. The problem then consists in rendering in a mathematically viable way a sequence of L amino acids.

---
[1] Some restrictions to the type of functions must be introduced, as the reader will see in the following.

Figure 3.1: Scheme of the vectorization process. The amino acid sequence of a unit peptide is translated into a mathematical entity, the sequence vector **x**.

Several ways to do this exist. Some of these consist in describing the amino acids by means of certain characteristics like polarity, acidity or basicity, hydropathy or even the size of the side chain, appropriately assembled in so-called "feature vectors". Others, like the one called here "standard representation" (see section 3.1.1) or the representations based on scoring matrices (BLOSUM [122] or PAM matrices [123]) require no knowledge of any physicochemical properties but similarly assign to each amino acid type a specific vector. Akin to the latter in its statistical nature, but a great deal more powerful, is the sequence profile [124–126] to be introduced in section 3.1.2. Like those found in BLOSUM or PAM matrices the entries of a sequence profile also represent amino acid affinities, but unlike those, do so in a *position-specific* fashion. For this reason profiles are sometimes called also position-specific scoring matrices (PSSM). The sequence profiles were originally employed for multiple sequence alignment. They provide a keener and far more "flexible" portrait of amino acid sequences because they contain a measure of the similarity between amino acid types as well as indirect global information about the protein chain these are inserted into, which is absent in other representations.

Throughout a large portion of the project, the standard sequence representation (the reason behind the "standard" attribute will become clear in the next section) was employed, and was replaced only lately, following what has become a driving trend in the secondary structure prediction community, by the profile representation. This greatly improved the program's performance.

The idea to use multiple sequence alignment as an aid in the secondary structure prediction quest dates back to the years between the late 1980s and the early 1990s [63, 127, 128] but the first program to employ sequence profiles in a systematic way was PHD [49], a set of feed-forward neural networks trained by back-propagation. Following the example of PHD, numerous other secondary structure prediction programs based on multiple sequence alignment and profiles [43, 51, 53, 55, 59, 61, 62, 67–69, 71, 77, 80–82, 85, 104] have been developed (see also section 1.1.3.2).

As a rule, both standard and profile representations of amino acids are based on a twenty-dimensional vector space, like the following examples will show. It is worth mentioning here though, that the project has at times also involved a partial dimensional reduction of the sequence vector space. This comes, as will be shown in section 4.2.2, as a consequence of grouping together, under a unique identifier, amino acids that are generally acknowledged to share certain physicochemical properties among each other and are therefore evolutionarily more likely to be interchanged across different proteins.

3.1.1 Standard representation

The standard representation straightforwardly assigns to each amino acid type a vector of the standard basis. So for instance

$$\mathbf{a}_{\text{Ala}} = \begin{pmatrix} 1 \\ 0 \\ 0 \\ 0 \\ \vdots \\ 0 \end{pmatrix}, \quad \mathbf{a}_{\text{Arg}} = \begin{pmatrix} 0 \\ 1 \\ 0 \\ 0 \\ \vdots \\ 0 \end{pmatrix}, \quad \mathbf{a}_{\text{Asn}} = \begin{pmatrix} 0 \\ 0 \\ 1 \\ 0 \\ \vdots \\ 0 \end{pmatrix}, \quad \ldots \quad \in \{0,1\}^{20}$$

and therefore a sequence comprising L amino acids results in the direct sum of L such vectors,

$$\mathbf{x}_{\text{std}} = \begin{pmatrix} \mathbf{a}_{A_1} \\ \mathbf{a}_{A_2} \\ \mathbf{a}_{A_3} \\ \vdots \\ \mathbf{a}_{A_L} \end{pmatrix} \in \{0,1\}^{L \times 20},$$

where $\mathbf{a}_{A_1}, \mathbf{a}_{A_2}, \mathbf{a}_{A_3}, \ldots, \mathbf{a}_{A_L}$ are the vectors corresponding to the amino acid types $A_1, A_2, A_3, \ldots, A_L$ which build up the sequence.

The standard representation has a limited correlation potential because it does not account for any similarities that may exist among different amino acid types and it does not contain any contextual information.

3.1.2 Profile representation

In the profile representation of a sequence, different vectors may be assigned to the same amino acid type, depending on the sequence, and indeed, on the protein chain context the amino acid finds itself in. While in the standard representation every amino acid vector $\mathbf{a}_{\text{xyz}}$ only has one non-zero entry and is always the same regardless to which sequence it belongs, in its profile counterpart all entries can be non-zero, each embodying the context-dependent affinity of the corresponding amino acid type.

A polypeptide of length L could be, for instance, translated into

$$\mathbf{x}_{\text{prof}} = \begin{pmatrix} \mathbf{p}_{A_1} \\ \mathbf{p}_{A_2} \\ \mathbf{p}_{A_3} \\ \vdots \\ \mathbf{p}_{A_L} \end{pmatrix} \in \mathbb{Z}^{L \times 20},$$

where

$$\mathbf{p}_{A_1} = \begin{pmatrix} -2 \\ +5 \\ +4 \\ -3 \\ \vdots \\ -4 \end{pmatrix}, \quad \mathbf{p}_{A_2} = \begin{pmatrix} +0 \\ -2 \\ -6 \\ -7 \\ \vdots \\ +1 \end{pmatrix}, \quad \mathbf{p}_{A_3} = \begin{pmatrix} -2 \\ +7 \\ -1 \\ -5 \\ \vdots \\ -1 \end{pmatrix}, \quad \ldots \in \mathbb{Z}^{20}.$$

The higher the vector entry, the greater the affinity of the corresponding amino acid type at the corresponding position.

3.2 Single-state likelihood functions

The so-called single-state scoring functions embody the correlation between the primary structure and the secondary structure pattern they represent the likelihood of. Since they are functions of the amino acid sequence, the sample space, that is to say the functions' definition space, will most reasonably be the vector space S of amino acid sequences introduced in section 3.1. Given a sample in S, the single-state likelihood functions will return a scalar value, which lies in an interval $[y_0, y_1]$, where y_0 stands for very unlikely and y_1 for very likely (or, less intuitively, the other way around).

In a distributed regression approach, the strength of a secondary structure character is extracted from each sequence in the learning data set: once properly normalized, this strength takes values in an interval very much like the one the likelihood functions take values in. Given the discrete nature of amino acid sequences, the strength "signals" coming from a residue-wise weighting of secondary structure patterns are not continuously distributed. Still, especially when considering longer sequence windows, their values can get fairly dense, so that suitable regression functions for the secondary structure under examination can be obtained through its interpolation.

In an exclusive regression approach it is not quite as plain to see how likelihood functions come about. In this case the assignment is clear-cut: either the sequence has got some secondary structure character or it has not (recall figure 2.8b). Such a sharp, non-continuous behaviour is not so obviously related to interval-valued functions like the ones sought. It is nevertheless still possible to treat the problem in the very same way, just with *two* possible strength levels only. Accordingly engineered exclusive regression functions, as the results will show, seem to perform slightly better than the respective distributed regression functions (see section 4.1).

3.2.1 Function optimization procedure

Independently from the approach being followed, the search for suitable likelihood functions puts the software up to the same task: find a function $f \in \mathcal{F}(S, \mathbb{R})$ that best fits the data in the learning set. The technique utilized here to find this function is that of straightforwardly minimizing the sum of the recognition squared errors.

3.2.1.1 Parametrical candidate estimators

The candidates to the role of single-state likelihood functions can in principle be chosen freely. In practice certain criteria need to be followed in order to enable a systematic implementation within a computer program. The functions' details are explored in the chapters of this report dedicated to the development of the predictor. Suffice it to say here that, since the space $\mathcal{F}(S, \mathbb{R})$ is quite large, it is convenient to restrict the choice to a feasible subset $\mathcal{G}_\mathbf{w}(S, \mathbb{R}) \subset \mathcal{F}(S, \mathbb{R})$ whose members have all the the same general shape but differ from each other in a number of free parameters grouped under the vector label $\mathbf{w}$. All candidates are, in other words, of the form $f(\mathbf{w}\,; \mathbf{x})$, where $\mathbf{w} \in \mathbb{R}^p$, $p \in \mathbb{N}$, are the parameters to be optimized to best adapt the function's behaviour to the learning data set.

This drastic but necessary restriction effectively precludes the chances of finding the "perfect" single-state likelihood functions (provided they exist at all). Even if the secondary structure prediction problem were exempt from the defects mentioned in chapter 2, enforcing the shape of the functions into so rigid paradigms introduces a gross approximation which renders the models hardly capable to cope with its complexity[2].

It is therefore very reasonable to assume that whatever candidate estimator is taken for the scoring function of interest, its action on a vector of the sample space S will be affected by a measure of "non-exactness". It is this "non-exactness" that is to be minimized in the optimization procedure, the result of which finally identifies the best candidate, among those available, for the secondary structure type under investigation.

---

[2]This is the reason why one should speak of function estimators, rather than plain functions, though the two concepts will be heedlessly confused from now on.

3.2.1.2 Least squares

The "non-exactness" of the estimators can be formalized by saying that, given a sequence vector $\mathbf{x} \in S$, its evaluation by means of the estimator f will, in general, be affected by an error ε, i.e.

$$y = f(\mathbf{w}\,;\mathbf{x}) + \varepsilon. \tag{3.2.1}$$

The *expected* value y spans different ranges depending on whether the distributed regression approach or the exclusive regression approach is followed. In the former, y carries the secondary structure patterns signals. In this case the range is therefore going to be the interval, let's write it $[\hat{y}_0, \hat{y}_1]$, which the signals' strengths have been constrained to take values in. In the latter, the two conditions, sequence pertaining or non-pertaining to the class, can be labeled with two numbers $\hat{y}_0$ and $\hat{y}_1$ and these values can in turn be assigned to y so the range is going to be, in this case, the set constituted by these very labels, i.e. $\{\hat{y}_0, \hat{y}_1\}$.

Let's assume now a set of data points is available, with which to instruct the not-yet-expert system. This will be, a learning set of pairs

$$D = \{(\mathbf{x}_i, y_i) \in S \times Y,\; i = 1, \ldots, N\},$$

where $Y = [\hat{y}_0, \hat{y}_1]$ for distributed regression, whereas $Y = \{\hat{y}_0, \hat{y}_1\}$ for exclusive regression, and suppose a family of scoring functions $f(\mathbf{w}) \in \mathcal{G}_{\mathbf{w}}(S, \mathbb{R})$, $\mathbf{w} \in \mathbb{R}^p$, $p \in \mathbb{N}$ contains the eligible estimators.

The parameter vectors $\mathbf{w}$ identify the members of the family and can be chosen freely once the shape of the functions has been decided. The ones sought are those $\mathbf{w}$ that minimize the norm

$$T(\mathbf{w}) = \|\underline{\varepsilon}(\mathbf{w})\|, \tag{3.2.2}$$

of the vector $\underline{\varepsilon} = (\varepsilon_1, \ldots, \varepsilon_N)$ of estimation errors on the set D.

Which norm is best to employ depends on several factors, like the quantity and the quality of the data points or the kind of functions that are being used to interpolate them. Some norms are mathematically better behaved than others, especially if the scoring functions are restricted to a convenient category, but this is not much of an issue in the context of computer applications in which disparate numerical methods allow to solve this kind of optimization problems by means of iterative approximations [129], quite regardless of the ill nature of the mathematical entities involved. It turns out however, that by appropriately selecting the kind of functions employed, not only it becomes possible to solve the problem analytically[3], but the choice of the norm itself is naturally induced too. The Gauss-Markov theorem [130] states in fact that *when estimating the parameters of a linear model* (that is to say the parameters appear linearly in the

---
[3] Or at least as analytically as a machine with limited computational power consents.

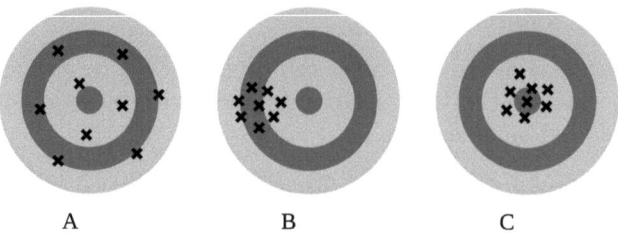

Figure 3.2: Intuitive illustration of different types of estimators by means of a dartboard example: candidate A represents an unbiased estimator because it hits uniformly around the target. The distribution of his darts shows that he is not very efficient though; candidate B is more efficient but biased to the left of the target. An efficient unbiased estimator here is represented by candidate C.

model) *in which the errors have zero expectation value, are uncorrelated and have equal variances, the most efficient unbiased linear estimators of the parameters themselves are the least-squares estimators, that is the estimators obtained by minimizing the standard 2-norm* $\|\underline{\varepsilon}(\mathbf{w})\|_2$ *of the error vector.*

Figure 3.2 exemplifies the concept of most efficient[4] unbiased estimator. Further encouraging the use of the 2-norm is the fact that, in case the errors follow a *normal distribution*, those obtained by minimizing 3.2.2 are the *maximum likelihood estimators* [131], that is, given the model of the system of interest, they are the estimators which *maximize* the likelihood of the observed data.

Throughout this project the theoretical advantages of the 2-norm

$$T(\mathbf{w}) = \|\underline{\varepsilon}(\mathbf{w})\|_2 \qquad (3.2.3)$$

were exploited and the first restriction in the spectrum of utilizable scoring functions was thus introduced. These were namely going to be *linear* in the parameters $\mathbf{w}$.

3.2.1.3 Linear equation system

Given these premises it can be shown [130] that the problem of minimizing (3.2.3), or rather its square

$$E(\mathbf{w}) = T^2(\mathbf{w}) = \sum_{i=1}^{N} \varepsilon_i^2(\mathbf{w}), \qquad (3.2.4)$$

where

$$\varepsilon_i(\mathbf{w}) \equiv f(\mathbf{x}_i; \mathbf{w}) - y_i \qquad (3.2.5)$$

---

[4]In rigorous mathematical terms this is the estimator with minimum variance.

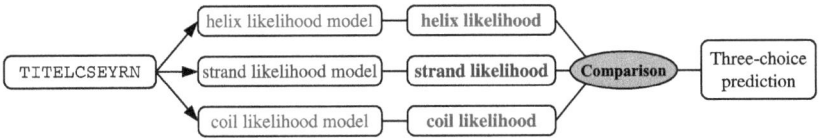

Figure 3.3: Example of multi-choice prediction with three secondary structure types. The three likelihood models return the likelihood values for helix, strand and coil. Typically, the residue of interest is the one at the central position in the sequence window.

is the error on data point i, boils down to solving a linear system of *normal equations* of the kind

$$A^T A\, \mathbf{w} = A^T\, \mathbf{b}_0, \qquad (3.2.6)$$

where the matrix A and the vector $\mathbf{b}_0$ solely depend on the data points $(\mathbf{x}_i, y_i)$ in the learning set D.

The system (3.2.6) always has solution and the function $f(\check{\mathbf{w}})$ with the parameters $\check{\mathbf{w}}$ that solve it, is the one among those of the chosen kind that best fits the data points of set D.

3.3 From single-state to multi-state

So far, it has been explained how to develop a tool, the single-state scoring function, that discerns a specific secondary structure pattern from the others. But secondary structure prediction is mostly regarded as a multi-state problem, typically as a three-state one (helix, strand, coil).

The optimization procedure, carried out for all secondary structure patterns of interest, provides a likelihood function for each of these,

$$f_\sigma : S \to \mathbb{R}, \quad \sigma \in \mathcal{P},$$

where $\mathcal{P}$ is the secondary structure patterns library, i.e. contains the indices identifying the secondary structure patterns. The functions f_σ typically take on values in ranges of the kind $(\hat{y}_0 - \delta_0, \hat{y}_1 + \delta_1)$, where δ_0 and δ_1 are comparatively small positive numbers, and it is reasonable to assume that the closer an amino acid sequence is mapped by f_σ to the "positive" label, the stronger is the hint that this sequence might present the secondary structure pattern σ. The idea is then simply to compare the values (see figure 3.3) and see which one is closest to its highest likelihood label. The secondary structure pattern it represents is then assigned to the sequence under examination.

This procedure assumes the different scoring functions to be equally renormalized, but this is true only to some extent. A certain difference between them exists, and represents another limitation of the method, but not such that it irremediably affects the performance of the predictor.

More refined techniques were actually also developed in the course of the project, that partially overcome the normalization problem. Since they were not generally applied though, their discussion is relegated to the proper sections (see section 5.2).

Part II
development

Chapter 4

Standard sequence-based predictor

4.1 Linear scoring functions

Probably the simplest function of the kind introduced in section 3.2 is a linear combination of the sequence vector components, x_k, $k = 1, \ldots, n$, where $n = L \times 20$ is the number of components, L being the length of the sequence window expressed in number of amino acids:

$$f(\mathbf{w};\mathbf{x}) = \mathbf{w} \cdot \mathbf{x}, \quad \mathbf{w} \in \mathbb{R}^n. \tag{4.1.1}$$

In this work the extension

$$f(\mathbf{w}, w_0;\mathbf{x}) = \mathbf{w} \cdot \mathbf{x} + w_0, \quad \mathbf{w} \in \mathbb{R}^n, \ w_0 \in \mathbb{R} \tag{4.1.2}$$

of (4.1.1) was used, where the offset w_0 appears beside the parameters $\mathbf{w}$. It is possible to reduce (4.1.2) back to the form (4.1.1) by introducing an extra variable x_0 which always takes on the value 1. This would however hide some interesting features of the least squares optimization procedure that are about to be illustrated.

Function optimization

The parameters $\mathbf{w}$ and w_0 are optimized following the criteria outlined in section 3.2.1.2. The function (3.2.4) is in this case[1]

$$E_S(\mathbf{w}, w_0) = \frac{1}{2N} \sum_{i=1}^{N} (\mathbf{w} \cdot \mathbf{x}_i + w_0 - y_i)^2, \tag{4.1.3}$$

---

[1] the reason for the additional factor $1/2N$ will soon be clear.

and the system of linear equations corresponding to (3.2.6) can be obtained by setting to zero all partial derivatives of (4.1.3) with respect to the parameters $\mathbf{w}$ and w_0. This results in the equations

$$0 = \frac{\partial E_S}{\partial w_k}(\mathbf{w}, w_0) = \frac{1}{N}\sum_{i=1}^{N} x_{ik}\left(\mathbf{w}\cdot\mathbf{x}_i + w_0 - y_i\right), \quad k=1,\ldots,n$$

and

$$0 = \frac{\partial E_S}{\partial w_0}(\mathbf{w}, w_0) = \frac{1}{N}\sum_{i=1}^{N}\left(\mathbf{w}\cdot\mathbf{x}_i + w_0 - y_i\right).$$

From the second equation, w_0 can be isolated,

$$w_0 = -\frac{1}{N}\sum_{i=1}^{N}\left(\mathbf{w}\cdot\mathbf{x}_i - y_i\right).$$

Subsequent insertion of this expression for w_0 into the first equations gives the linear equations system

$$0 = \frac{1}{N}\sum_{i=1}^{N} x_{ik}\left(\mathbf{w}\cdot\mathbf{x}_i - \frac{1}{N}\sum_{j=1}^{N}(\mathbf{w}\cdot\mathbf{x}_j - y_j) - y_i\right), \quad k=1,\ldots,n,$$

where the index of the internal sum has been renamed. Using the full expression for the scalar products and the definition of mean over a sample, the equations become for $k = 1, \ldots, n$

$$\begin{aligned}
0 &= \frac{1}{N}\sum_{i=1}^{N} x_{ik}\left(\sum_{h=1}^{n} w_h x_{ih} - \frac{1}{N}\sum_{j=1}^{N}\left(\sum_{h=1}^{n} w_h x_{jh} - y_j\right) - y_i\right) \\
&= \frac{1}{N}\sum_{i=1}^{N} x_{ik}\left(\sum_{h=1}^{n} w_h x_{ih} - \sum_{h=1}^{n} w_h \frac{1}{N}\sum_{j=1}^{N} x_{jh} + \frac{1}{N}\sum_{j=1}^{N} y_j - y_i\right) \\
&= \frac{1}{N}\sum_{i=1}^{N} x_{ik}\left(\sum_{h=1}^{n} w_h x_{ih} - \sum_{h=1}^{n} w_h \langle x_h\rangle + \langle y\rangle - y_i\right) \\
&= \sum_{h=1}^{n}\left(\langle x_k x_h\rangle - \langle x_k\rangle\langle x_h\rangle\right) w_h + \langle x_k\rangle\langle y\rangle - \langle x_k y\rangle \\
&= \sum_{h=1}^{n} \mathrm{cov}(x_k, x_h) w_h - \mathrm{cov}(x_k, y).
\end{aligned}$$

The equations above can be written in the matrix form

$$\Gamma_S \mathbf{w} = \mathbf{b}_S, \tag{4.1.4}$$

where $\Gamma_S = \text{cov}(\mathbf{x}, \mathbf{x})$ is the *sequence covariance matrix*, i.e. the matrix whose elements span the covariances $\text{cov}(x_k, x_h)$, for $h, k = 1, \ldots, n$, and $\mathbf{b}_S$ is the vector whose components are the covariances $\text{cov}(x_k, y)$ for $k = 1, \ldots, n$, between the sequence and the secondary structure.

Regularization

In reality, in place of equation (4.1.4) its *regularized* version

$$(\Gamma_S + \lambda) \mathbf{w} = \mathbf{b}_S, \tag{4.1.5}$$

was employed, which is obtained by adding a regularization term to the function (4.1.3), so that this actually reads

$$E_S(\mathbf{w}, w_0, \lambda) = \frac{1}{2N} \sum_{i=1}^{N} (f(\mathbf{w}, w_0; \mathbf{x}_i) - y_i)^2 + \lambda \mathbf{w} \cdot \mathbf{w}. \tag{4.1.6}$$

The regularization term serves the double purpose of removing possible near-singular behaviours and of keeping the size of the parameters $\mathbf{w}$ small to attenuate their overfitting to the sequence samples contained in the learning data set. It therefore plays an especially important role when the number of parameters available is large[2] as compared to the number of samples used for the function's optimization procedure. As will soon be shown, this mostly is not the case for linear scoring functions like (4.1.2), so that a more thorough discussion of the subject is best postponed to later sections (see for instance section 5.2.3 and the test results of section 5.2.4). Until further notice, it is to be assumed a value $\lambda = 10^{-5}$ was used.

Test results

Unit peptides of increasing lengths were isolated from the designated portion of the $\mathcal{R}_{50/50}$ subdivisions to fuel independent learning processes for helix, strand and coil. Only fully defined peptides, that is to say with no missing residues, were employed (as the reader shall see in section 4.1.5.1, this requirement would be dropped later on, since suspected to lead to certain biases in the prediction performance). The true secondary structure strings were assembled applying the loose reduction scheme (EVA's standard of truth [96]) to the eight DSSP states (see section 2.3.1.5).

The scoring functions were put to test on both the validation subset and the learned subset in order to have a measure of their generalization capability. In the following, the term *prediction* shall preferably be reserved for the test carried

---
[2]The meaning of large actually depends on the nature of the data and of the scoring functions used.

out on the validation subset whereas the test carried out on the learned subset will be referred to as *recognition*.

The Q_3 prediction and recognition accuracies of predictors constructed upon four different types of scoring functions were compared. The scoring functions comprised exclusive regression functions (ESF) and three types of distributed regression[3] functions (DSF) differing in the way the secondary structure motifs at all positions along the sequence were weighed to get the regression signals:

exclusive (ESF): the central amino acid position is the key-position and gets weight one; all others get no weight at all;

gaussian distributed (gDSF): the secondary structure weights follow a gaussian distribution centred in the central amino acid;

binomially distributed (bDSF): the secondary structure weights follow a binomial distribution centred in the central amino acid;

uniformly distributed (uDSF): all motifs have the same weight.

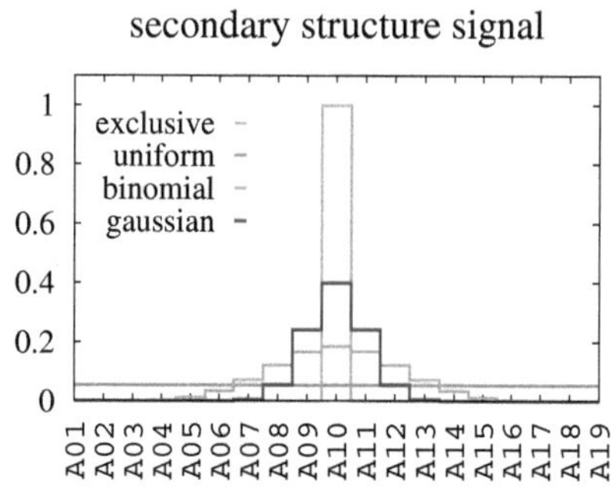

Figure 4.1: Example of the different secondary structure signal distributions in a sequence window comprising 19 amino acids. All distributions are normalized in such a way that the sum of all signals equals 1. In the abscissa are the 19 amino acid positions. The gaussian distribution is the one corresponding to $\delta^2 = 2$.

Figure 4.1 gives a visual idea of how the three different distributions work and compare to the exclusive regression approach. Incidentally, ESFs can be regarded as the limit case of gDSFs were the weights distribution tends to the normalized *delta*-function

$$\Delta(x-c) = \begin{cases} 1 & \text{for } x = c \\ 0 & \text{elsewhere,} \end{cases}$$

where c is the central position in the sequence window.

Figure 4.2 contains the prediction and recognition performances achieved on the $\mathcal{R}_{50/50}$ data set subdivisions (see section 2.3.2.2s) using scoring functions based on sequence windows of different lengths. Notice that only windows spanning an uneven number of amino acids were taken into account. As can be seen, an increase of the window length corresponds initially to an increase, but subsequently to a decrease in performance in both prediction and recognition. Two

---

[3]Though the tests on distributed regression scoring functions were carried out only later in the course of the project, it is perhaps best to analyse them here in direct comparison with those carried out on the original exclusive regression scoring functions.

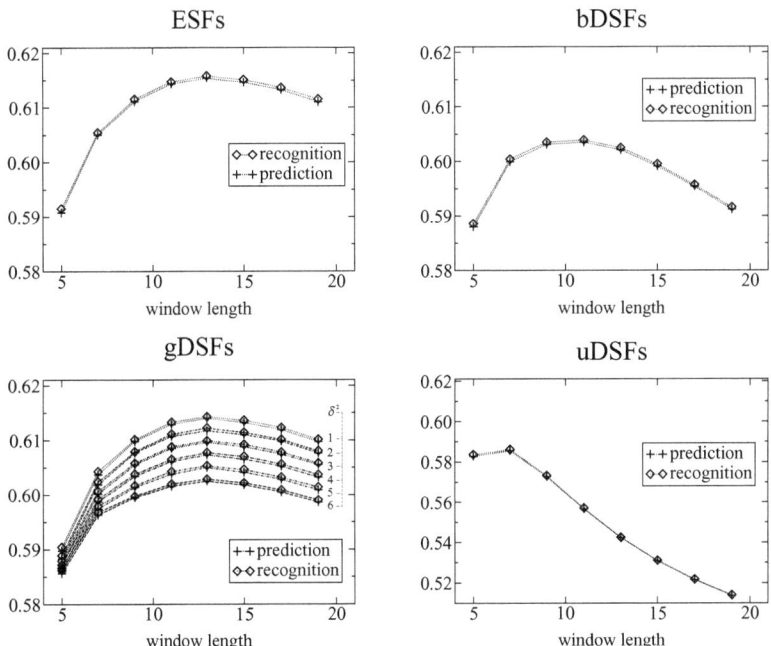

Figure 4.2: Q_3-accuracy averages (loose DSSP reduction scheme) for recognition and prediction on the $\mathcal{R}_{50/50}$ data set subdivisions using linear scoring functions based on sequence windows of different lengths. The $\bar{Q}_3$ achieved by gDSFs clearly approaches the one achieved by ESFs when δ^2 gets smaller. [Note that the Q_3-scale in the plot relative to uDSFs is different from that of the other plots.]

possible causes were identified, which could explain the observed behaviour. The first was merely of technical nature and had to do with the fact that while only fully defined peptides were learned, a prediction was actually performed for all residues, including those at the N- and C-termini. The amount of *missing* sequence information at the N- and C-termini increases with the window length, so that predictors based on larger sequence windows suffer a comparatively larger handicap. The second cause envisaged the possibility that the secondary structure motif of a residue could actually, in general, be badly correlated with information about residues too far away along the sequence. This would rather represent noise, the only effect of which would be that of diluting the useful contributions coming from the residues at the positions that most (or exclusively) contribute to the secondary structure signal. In the case of uDSFs, the distant residues, weighed as much as the central ones, apparently scramble the secondary structure signals already in sequence windows of nine amino acids. Subsequent tests (see section 4.1.5.2), in which this behaviour was no longer observed, would indicate that the first cause, i.e. the technical artifact, was probably to be held accountable for the most part of it.

From table 4.1 and visually from figure 4.2 itself, it is also possible to see how, as expected, the prediction performance systematically lies a fraction of a percent below the recognition performance, the gap between the two widening a little as the window lengths increase. This means that the sequences learned are represented slightly better by the parameters of the scoring functions. This is especially true for the parameters relative to the more distant residues, which are otherwise, as earlier noticed, only loosely correlated with the central secondary structure motifs. This effect is however merely physiological and rather weak so that one may not speak of overfitting. Another remark to make is that ESFs were awarded, in general, a higher Q_3 index than DSFs.

Given the state of the art of secondary structure prediction as known from the introduction, even the best Q_3-accuracy ratio results of about 0.615 obtained by the ESFs based on sequence windows of 13 amino acids are not even close to those of other methods. In the following, some of the upgrades devised to reduce the gap will be analysed.

4.1.1 Dual statistics

The first upgrade which was worked out involved a modification of the objective function (4.1.3) to account for the difference between the so-called *positive* and *negative* samples.

The optimization procedure of the exclusive regression functions is naturally suited to treat positive and negative samples separately: positive would be all those samples whose central secondary structure motif matched the one of interest for the scoring function under development, negative the others.

Table 4.1: Q_3-accuracy averages (loose DSSP reduction scheme) for recognition and prediction on the $\mathcal{R}_{50/50}$ data set subdivisions using linear scoring functions based on sequence windows of different lengths.

L	$n+1$	$\bar{Q}_3$ **pred.**	$\bar{Q}_3$ **recog.**	$\bar{Q}_3$ **pred.**	$\bar{Q}_3$ **recog.**
		ESFs		**bDSFs**	
5	101	0.591 ± 0.001	0.591 ± 0.001	0.588 ± 0.001	0.589 ± 0.001
7	141	0.605 ± 0.001	0.605 ± 0.001	0.600 ± 0.001	0.600 ± 0.001
9	181	0.611 ± 0.001	0.611 ± 0.001	0.603 ± 0.001	0.603 ± 0.001
11	221	0.614 ± 0.001	0.615 ± 0.001	0.603 ± 0.001	0.604 ± 0.001
13	261	0.615 ± 0.001	0.616 ± 0.001	0.602 ± 0.001	0.602 ± 0.001
15	301	0.615 ± 0.001	0.615 ± 0.001	0.599 ± 0.001	0.599 ± 0.001
17	341	0.613 ± 0.001	0.614 ± 0.001	0.595 ± 0.001	0.596 ± 0.001
19	381	0.611 ± 0.001	0.611 ± 0.001	0.591 ± 0.001	0.592 ± 0.001
		gDSFs ($\delta^2 = 1$)		**uDSFs**	
5	101	0.590 ± 0.001	0.590 ± 0.001	0.583 ± 0.001	0.584 ± 0.001
7	141	0.604 ± 0.001	0.604 ± 0.001	0.585 ± 0.001	0.586 ± 0.001
9	181	0.610 ± 0.001	0.610 ± 0.001	0.573 ± 0.001	0.573 ± 0.001
11	221	0.613 ± 0.001	0.613 ± 0.001	0.557 ± 0.001	0.557 ± 0.001
13	261	0.614 ± 0.001	0.614 ± 0.001	0.542 ± 0.001	0.542 ± 0.001
15	301	0.613 ± 0.001	0.614 ± 0.001	0.531 ± 0.001	0.531 ± 0.001
17	341	0.612 ± 0.001	0.612 ± 0.001	0.521 ± 0.001	0.522 ± 0.001
19	381	0.610 ± 0.001	0.610 ± 0.001	0.514 ± 0.001	0.514 ± 0.001

Finding a meaningful subdivision of the data samples for the optimization of distributed regression functions on the other hand, is a little less straightforward: the way chosen was to set a score threshold for every motif and distinguish the samples whose score lay above this threshold from those whose score lay below it. The score thresholds were tuned following a simple heuristic procedure.

The function (4.1.3) was replaced by

$$E_D = \frac{1}{4N^+} \sum_{i=1}^{N^+} (f(\mathbf{x}_i) - y_i)^2 + \frac{1}{4N^-} \sum_{j=1}^{N^-} (f(\mathbf{x}_j) - y_j)^2. \quad (4.1.7)$$

Here N^+ and N^- are the number of positive and negative samples respectively. The sums appearing in (4.1.7) are understood to run respectively on the subset of positive and negative samples.

Given E_D, the very same steps followed in the previous section can be repeated to get the final linear equations system. This turns out to be formally identical to that of (4.1.4), with the difference that the averages appearing in the covariances are weighted averages on positive and negative samples. More concretely it is now,

$$\Gamma_D \mathbf{w} = \mathbf{b}_D, \quad (4.1.8)$$

where

$$[\Gamma_D]_{kh} = \langle x_k x_h \rangle_\pm - \langle x_k \rangle_\pm \langle x_h \rangle_\pm$$

$$b_{Dk} = \langle x_k \rangle_\pm \langle y \rangle_\pm - \langle x_k y \rangle_\pm,$$

and, in general,

$$\langle q \rangle_\pm = \frac{1}{2N^+} \sum_{i=1}^{N^+} q_i + \frac{1}{2N^-} \sum_{j=1}^{N^-} q_j = \frac{w^{(0)}}{N^+} \sum_{i=1}^{N^+} q_i + \frac{w^{(0)}}{N^-} \sum_{j=1}^{N^-} q_j, \qquad (4.1.9)$$

the sums being once again understood as partial sums over positive and negative subsets. The factor $w^{(0)} = 1/2$, stemming from (4.1.7), is necessary to preserve the normalization of the averages.

Contrary to the expectations, the separation of positive and negative samples in the statistics lead to a sensible decrease in performance for ESFs, which even fell some measure behind bDSFs and gDSFs. This outcome could be related to the fact that the Q_3 index evaluates the quality of the prediction for the three main secondary structure states and could, to a certain extent, overshadow upgrades that are mainly devised to improve the recognition of a *single* secondary structure state. Partial results, relative to single-state prediction will be presented in later sections (starting with section 4.1.5), where these issue will be further investigated.

Dual statistics introduce at any rate an additional degree of freedom that was yet to be exploited: they enable to *reweight* positive and negative sums at will. This is accomplished by changing (4.1.7) to

$$E_D^* = \frac{w^+}{2N^+} \sum_{i=1}^{N^+} (f(\mathbf{x}_i) - y_i)^2 + \frac{w^-}{2N^-} \sum_{j=1}^{N^-} (f(\mathbf{x}_j) - y_j)^2, \qquad (4.1.10)$$

where alongside the function parameters **w** also the reweighting factors w^+ and w^- appear, incidentally causing the redefinition of the weighted average (4.1.9)

$$\langle q \rangle_\pm = \frac{w^+}{N^+} \sum_{i=1}^{N^+} q_i + \frac{w^-}{N^-} \sum_{j=1}^{N^-} q_j. \qquad (4.1.11)$$

The same issues mentioned earlier about preserving the normalization of the averages enforce here the condition that w^+ and w^- be themselves normalized, that is, that $w^+ + w^- = 1$, so that effectively one of the two is determined once the other is assigned.

Calling Γ_D^* and b_D^* the new covariance matrix and vector respectively, the optimization of (4.1.10) results in the linear equations system

$$\Gamma_D^* \mathbf{w} = \mathbf{b}_D^*. \qquad (4.1.12)$$

The function E_D^* depends non-linearly on the reweighting parameters. This means they can't be included in the squared errors minimization procedure described earlier. A simple heuristic approach was followed here, identical to the one used to optimize the score thresholds, to find the "closest to optimal" w^+s *simultaneously* for all secondary structure motifs.

Table 4.2: Q_3-accuracy (loose DSSP reduction scheme) for recognition and prediction on the $\mathcal{R}_{50/50}$ data set subdivisions using linear scoring functions based on a sequence window of 13 amino acids. These were optimized in turn with non-reweighted dual statistics and dual statistics with heuristically tuned reweighting of positive and negative partial sums. For DSFs only the results obtained with the optimal score thresholds are reported.

dual statistics	$\bar{Q}_3$ **pred.**	$\bar{Q}_3$ **recog.**	$\bar{Q}_3$ **pred.**	$\bar{Q}_3$ **recog.**
	ESFs		bDSFs	
non-reweighted	0.605 ± 0.001	0.605 ± 0.001	0.607 ± 0.001	0.608 ± 0.001
best reweighted	0.617 ± 0.001	0.617 ± 0.001	0.608 ± 0.001	0.608 ± 0.001
	gDSFs, $\delta^2 = 1$		uDSFs	
non-reweighted	0.615 ± 0.001	0.616 ± 0.001	0.541 ± 0.001	0.542 ± 0.001
best reweighted	0.615 ± 0.001	0.616 ± 0.001	0.543 ± 0.001	0.543 ± 0.001

Test Results

The results obtained using dual statistics, both with and without reweighting, are reported in table 4.2. For DSFs, only the results obtained using heuristically optimized score thresholds are reported. The scoring functions were based on the optimal sequence window of 13 amino acids.

Interestingly, only ESFs seemed to appreciably benefit from the reweighting, the best positive reweighting factors being $w^+ = 0.6$ for helix, $w^+ = 0.5$ for strand and $w^+ = 0.7$ for coil. Despite the recovery they enjoyed however, a real breakthrough lay still ahead. The improvement on unified statistics, of the same order of magnitude of the deviation, hardly constituted ground for jubilation.

Is it actually learning? In order to rule out the possibility that bugs of the program might be drastically affecting its performance, a "controlled overfitting" test was performed. The size of the learning data sets was progressively reduced to see if the recognition accuracy was thereby augmented. The accuracy on a fixed validation data set comprising 700 protein domains was monitored.

The test was carried out using ESFs based on the optimal sequence window of 13 amino acids. The functions were optimized with dual statistics, but using no reweighting, since it is reasonable to assume that the optimal reweighting factors strongly depend on the populations of the secondary structure motifs, and these could sensibly fluctuate among the random subdivisions, especially when few samples were learned. As can be seen in figure 4.3, the smaller the number of unit peptides the program learned the worse its prediction capability became. While this was progressively impaired and eventually totally compromised however, its recognition capability kept improving. When the size of the learning data set was smaller than the number of parameters ($N \lesssim n = 260$) the learned samples were recognized with 100% accuracy ($Q_3 = 1.0$). This excluded the presence of major bugs.

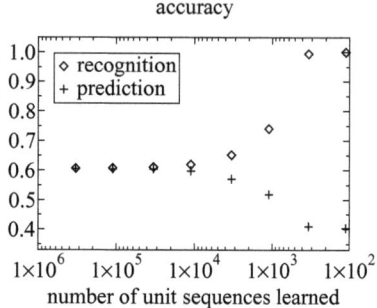

Figure 4.3: Average prediction and recognition Q_3-accuracy (loose DSSP reduction scheme) for different learning data set sizes. The number of unit sequences learned is reported in reverse logarithmic scale in abscissa. The deviations are not shown. For large learning data sets the recognition and prediction rates come very close. As the number of unit sequences learned decreases the recognition rate grows to reach 100% while the prediction rate approaches that of a random guess.

4.1.2 Filtering the learning data set

A technique investigated at relatively early stages in the development of the program, was that of *filtering* the learning data sets. The rationale behind this idea was that having the scoring functions learn *bad* samples could affect their overall prediction capability. This would then benefit from a partial removal of the former from the original learning data set.

4.1.2.1 Positive reinforcement

The first filtering system experimented with involved a *duplication* of the learning/recognition procedure (see figure 4.4). In the duplicate procedure, the learning data set was restricted to those samples the secondary structure of which was properly recognized, or *nearly* so, in the first procedure.

Stage-I of the procedure was carried out in the traditional way, only monitoring the performance on the learning data set. Since it was desirable to be able to calibrate the filter so that the learning data set could be flexibly readjusted, two tunable parameters were introduced, the safety threshold t_s and the criticality threshold t_c, to determine whether a prediction was wrong *enough*, as described in the following.

For all those residues whose motif was not recognized, the secondary structure likelihood scores were examined. The sequence sample corresponding to that residue was discarded in the next round of learning if at least one of the following conditions was observed:

- the likelihood score for the actual secondary structure was below t_s;
- the maximum likelihood was above t_c;

To test this, scoring functions based on a sequence length of 13 amino acids were employed. These were optimized using dual statistics and no reweighting of positive and negative sums (see arguments on page 53). To simplify the problem,

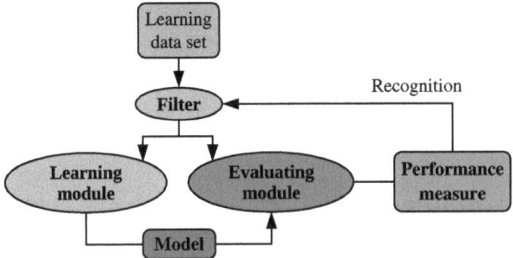

Figure 4.4: Scheme of the positive reinforcement technique. A first recognition procedure serves to identify potentially difficult sequences. These are filtered out of the learning data set in the second learning procedure.

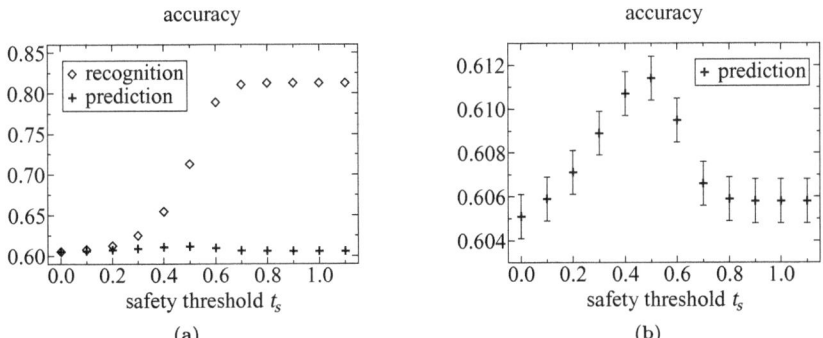

Figure 4.5: Q_3-accuracy averages (loose DSSP reduction scheme) (a) in recognition and prediction and (b) focused on prediction only, for different data set readjustment filters. All scoring functions were based on a sequence window of 13 amino acids and optimized using non-reweighted dual statistics. The strength of the filters is represented here by t_s ($t_c = 1 - t_s$).

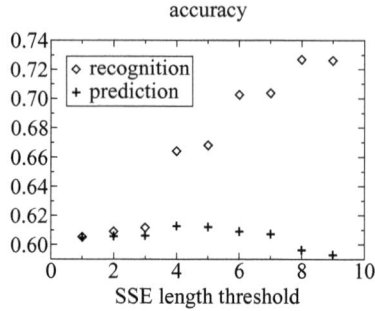

Figure 4.6: Q_3-accuracy values in recognition and prediction (loose DSSP reduction scheme), using progressively larger SSE length threshold filters. All scoring functions were based on a window of 13 amino acids and optimized using nonreweighted dual statistics. An interesting feature of the plot is the stepwise dependence on the length thresholds.

the filter parameters were constrained according to

$$t_s = 1 - t_c. \qquad (4.1.13)$$

The Q_3-accuracy results obtained on the $\mathcal{R}_{50/50}$ subdivisions are displayed in figure 4.5. As can be better seen in the close-up of figure 4.5b, reducing the learning data set to "easier" samples does bring an improvement, though it is as usual better not to restrict it too much lest run into overfitting. This can already be observed at the best performing filtering level ($Q_3 \approx 0.6114$), corresponding to $t_s = t_c = 0.5$. Clearly, the fact that "difficult" samples had been expelled from the learning data sets, however not from the validation data sets, dramatically increased the gap between the performances on the two. Though not unsuccessful, due to the complication it introduced of duplicating the learning procedure, this two-stages technique was put aside in the development process thereafter.

4.1.2.2 SSE length filters

Another filtering system tested was the one based on the lengths of the secondary structure elements (SSEs).

The idea consisted in removing from the learning data set all unit sequences in which the secondary structure key belonged to *short* SSEs, that is, SSEs involving only *few* amino acids. Encouraging the design of this scheme was also the consideration that short SSEs can often be artifacts of the secondary structure definition. The only detail left open was the thresholds below which SSEs were to be deemed short. First of all, only helix and strand classes were affected by the length filters. No constraints were set on the length of coil segments, since these often involve few residues only (e.g. tight turns). To further simplify the procedure, the same length threshold was applied to helix and strand SSEs. SSE length thresholds ranging from one to nine amino acids were tested, one effectively meaning "no threshold". The functions used were based on sequence windows comprising 13 amino acids and optimized using dual statistics. Following again the arguments of page 53, no reweighting of positive and negative samples was applied. The Q_3-accuracy results on the $\mathcal{R}_{50/50}$ subdivisions are collected in figure 4.6. As the results show, it does seem to pay off ($Q_3 \approx$ 0.6126, i.e. three quarters of a percent better than with no length thresholds) to exclude the shortest SSEs from the learning procedure, the optimal threshold corresponding to four residues. To ensure the filters to be actually the most convenient, similar tests were carried out abandoning the requirement that the filter be the same for both helix and strand classes: the results still spoke in favour of using a common length threshold of four residues.

Due to its relative efficacy, the SSE length filtering technique kept being employed in some applications involving linear scoring functions. It was not given much space however in the subsequent development. As will be shown later on in fact, the most effective predictors turned out to be the ones based on quadratic scoring functions, which depend on a considerably larger number of parameters. For such predictors a reduction of the learning set is not advisable.

4.1.3 The secondary structure key-position

A quite crucial issue in the optimization of exclusive regression scoring functions is the *selection* of the secondary structure information.

All ESFs seen so far were constructed by correlating the sequence samples to the secondary structure motif of the central residue. While this probably *is* the most reasonable choice, it is unclear to what extent such a simple approach makes use of the information available. This consideration motivated the design of a mechanism that would allow to better exploit the sequence-structure correlations offered by the data. This mechanism consisted in independently optimizing not one but several scoring functions for all secondary structure motifs of interest. For each of these, the amino acid sequence would be correlated to the secondary structure motif at a *different* position (see figure 4.7). This way, if each scoring function was still aware of the secondary structure motif at one position only, all of them together were somehow providing a more complete picture of the secondary structure landscape of the peptide. The contributions from all these scoring functions were merged in the final test phase as shown in figure 4.8. The essential part of the procedure, involving how to actually put all the scores together, consisted in quite naively averaging the available contributions, with the only condition that these be given different weights, according to the scheme to be illustrated shortly.

As can be seen from figure 4.9, if taken as the only source of secondary structure information, the motifs inherent to the various positions in a unit peptide provide qualitatively different correlations with the amino acid sequence. In particular, as is to be expected, the internal positions correlate better than the peripheral ones. However, the closer it gets to the centre the smaller a difference it seems to make which position is actually considered (see results for

Figure 4.7: Example of a learning scenario involving more than one key-position. The same unit peptide is learned by different scoring functions but is assigned in each case the secondary structure of a residue at a different position.

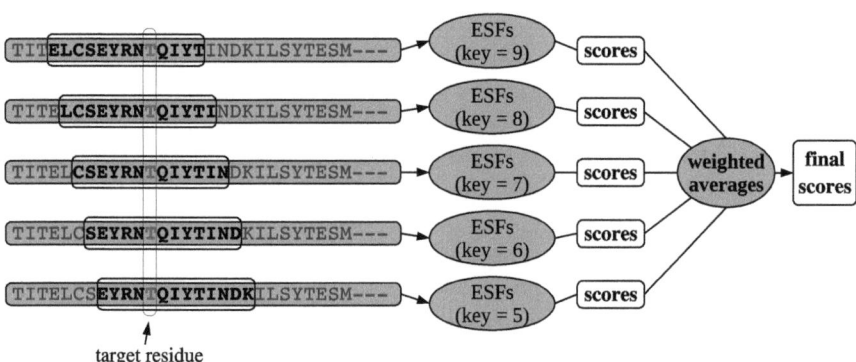

Figure 4.8: Illustrative view of the use of scoring functions optimized to correlate the sequence to the secondary structure motifs at five different key-positions. In this example the key-positions employed are the five innermost in a sequence window comprising 13 residues. In order to get the likelihoods for the threonine residue (cerise), the five sequence windows having that residue at the five key-positions are extracted and passed to the five scoring functions. The five resulting scores are then weighed and averaged to get the final predicted likelihoods.

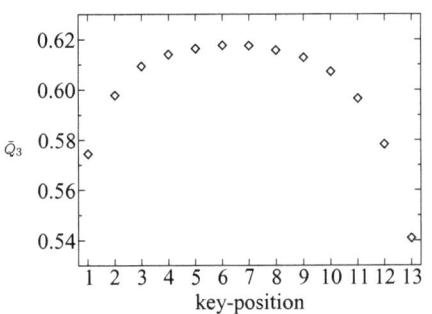

key	$\bar{Q}_3$ recog.
1	0.5746 ± 0.0010
2	0.5979 ± 0.0009
3	0.6095 ± 0.0007
4	0.6142 ± 0.0008
5	0.6166 ± 0.0009
6	0.6176 ± 0.0009
7	0.6175 ± 0.0009
8	0.6159 ± 0.0009
9	0.6129 ± 0.0010
10	0.6073 ± 0.0009
11	0.5967 ± 0.0009
12	0.5785 ± 0.0009
13	0.5411 ± 0.0010

Figure 4.9: Average Q_3-accuracy (loose DSSP reduction scheme) in recognition for linear scoring functions based on sequence windows with 13 amino acids using different secondary structure key-positions. The optimization was carried out using dual statistics with positive/negative reweighting factors $w_{\text{helix}}^+ = 0.6$, $w_{\text{strand}}^+ = 0.5$, and $w_{\text{coil}}^+ = 0.7$. No SSE length filters were applied. The averages are computed using (2.3.1) on the $\mathcal{R}_{50/50}$ data set subdivisions. The deviations, computed according to (2.3.2), are not reported. Clearly visible is the asymmetry between N-terminus and C-terminus. Scoring functions optimized using the inner key-positions show similar performances. When very peripheral key-positions are taken, it makes a considerable difference whether they are taken close to the N-terminus or to the C-terminus. In particular, the secondary structure motifs at the N-terminus of a peptide seem to be better correlated with the sequence than their counterparts at the C-terminus.

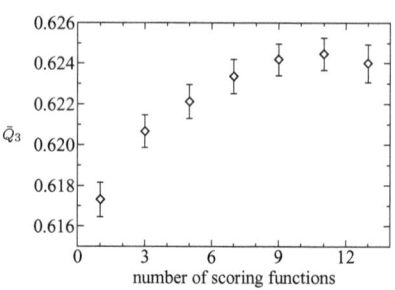

Figure 4.10: Average recognition Q_3-accuracy (using loose DSSP reduction scheme), using up to 13 scoring functions with different key positions for each secondary structure type. In all cases, the innermost key-positions were used. [See figure 4.9 for further details on the test's background.]

key-positions '6' and '7'). The Q_3-accuracy gives a measure of the predictor's reliability. It is therefore possible to take the average recognition Q_3-accuracies from figure 4.9 to weigh the contributions of different scoring functions to the average likelihood scores.

The next issue to be addressed concerned *how many* of the available scoring functions should actually contribute to the averages. While in fact all of them could be used in principle, the suspicion was it could be advisable to leave out those corresponding to the most peripheral key-positions, since these, as suggested by figure 4.9, present a rather poor correlation with the amino acid sequence.

To find the best possible scoring function combination, a series of tests were carried out. These involved using increasingly larger score sets obtained by gradually introducing more peripheral key-positions.

The recognition Q_3-accuracy results in figure 4.10 provide a clear indication that it is rewarding to use nearly all scores, leaving out only those stemming from the very terminal key-positions. The motifs at these positions are apparently too loosely correlated to the amino acid sequence and do not provide any useful signal to the global scores. The best results for recognition and prediction, relative to average scores built up out of eleven key-specific scores, are collected in table 4.3. Though still far from competitive, the average prediction Q_3 appeared to validate the key-position averaging illustrated and justify its subsequent development. At a later stage (see section 4.1.5.1) though, doubts would be raised about its efficacy, to the point that it would eventually be dropped.

4.1.4 The secondary structure motifs unleashed

Considering the secondary structure motifs at more than one key-position is one way to extend the sequence-structure correlation information. Another way is to introduce and optimize scoring functions for *more* secondary structure types than the ones of interest. The contributions of such scoring functions are filtered in the prediction phase by applying appropriate reduction criteria. The following example will clarify the idea.

Table 4.3: Average Q_3-accuracy for prediction and recognition on the $\mathcal{R}_{50/50}$ data set subdivisions. For each secondary structure type of interest a set of eleven scoring functions was employed. These were based on sequence windows of 13 amino acids and optimized, using dual statistics with heuristically tuned weights, to correlate to the secondary structure of the eleven innermost key-positions.

		prediction		recognition	
	DSSP	Sens_3	Spec_3	Sens_3	Spec_3
helix	[H,G,I]	0.6099 ± 0.0021	0.6475 ± 0.0028	0.6106 ± 0.0012	0.6492 ± 0.0021
strand	[E,B]	0.3893 ± 0.0018	0.5660 ± 0.0048	0.3892 ± 0.0024	0.5656 ± 0.0030
coil	[C,S,T]	0.7621 ± 0.0019	0.6251 ± 0.0010	0.7625 ± 0.0008	0.6249 ± 0.0013
$\bar{Q}_3^{(\text{loose})}$		0.6237 ± 0.0008		0.6242 ± 0.0008	

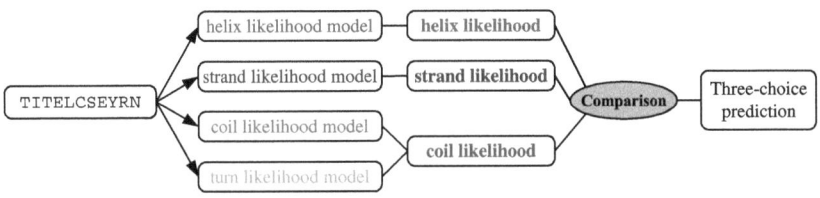

Figure 4.11: Example of prediction scenario with four secondary structure likelihood models. The likelihood signal for the coil meta-class is obtained here by taking the maximum among the likelihoods of the turn and the true coil classes.

Example: the hydrogen-bonded turn. In the minimal extension of the basic secondary structure alphabet (see figure 4.11) the very populated and heterogeneous coil class was divided in two subclasses: turn and true coil. The scoring function optimization procedure was then applied to both subclasses so that in the end the scoring functions to be compared to each other were four instead of three. In the testing phase an amino acid was assigned the general coil class if either of the two scoring functions, the one giving the likelihood of true coil or the one giving the likelihood of turn, returned the strongest signal.

Table 4.5 displays the average Q_3 and partial accuracy (motif-specificities, motif-sensitivities) results obtained by exploiting progressively higher levels of detail of the DSSP and STRIDE secondary structure definitions (see table 4.4). All scoring functions were based on a sequence window of 13 residues. Due to the larger number of motifs involved here, the heuristic procedure used earlier to optimize the dual statistics reweighting factors would have become impractical. The use of non-reweighted dual statistics, on the other hand, would have introduced a bias, due to the considerable variations among the populations of different secondary structure types. Therefore, unified statistics were preferred to dual statistics. No SSE length filters were applied.

Table 4.4: Levels of secondary structure definition. Levels 1 to 6 refer to pure DSSP (or STRIDE) assignments. Levels 7 to 11 arise from a further reclassification of five DSSP motifs ("H", "E", "G", "I", and coil) in "short" and "long", depending on the length (in terms of residues) of the SSE they belong to: length thresholds of seven, four and three residues were used for helix, strand and coil respectively. Levels 12 to 15 encompass the distinction of motifs belonging to long helix- and strand-like SSEs according to the residue's relative position within the SSE itself.

secondary structure	definition level														
	1	2	3	4	5	6	7	8	9	10	11	12	13	14	15
helix/α-helix	✓	✓	✓	✓	✓	✓	✓	✓	✓	✓	✓	✓	✓	✓	✓
N-term. α-helix												✓	✓	✓	✓
C-term. α-helix												✓	✓	✓	✓
short helix/α-helix								✓	✓	✓	✓	✓	✓	✓	✓
strand/β-strand	✓	✓	✓	✓	✓	✓	✓	✓	✓	✓	✓	✓	✓	✓	✓
N-term. β-strand												✓	✓	✓	✓
C-term. β-strand												✓	✓	✓	✓
short strand/β-strand								✓	✓	✓	✓	✓	✓	✓	✓
3-turn-helix				✓	✓	✓	✓	✓	✓	✓	✓	✓	✓	✓	✓
N-term. 3-turn-helix												✓	✓	✓	✓
C-term. 3-turn-helix												✓	✓	✓	✓
short 3-turn-helix								✓	✓	✓	✓	✓	✓	✓	✓
5-turn-helix				✓	✓	✓	✓	✓	✓	✓	✓	✓	✓	✓	✓
N-term. 5-turn-helix												✓	✓	✓	✓
C-term. 5-turn-helix												✓	✓	✓	✓
short 5-turn-helix								✓	✓	✓	✓	✓	✓	✓	✓
β-bridge				✓	✓	✓	✓	✓	✓	✓	✓	✓	✓	✓	✓
bend			✓			✓	✓			✓	✓			✓	✓
turn		✓	✓		✓	✓		✓	✓	✓			✓	✓	✓
coil	✓	✓	✓	✓	✓	✓	✓	✓	✓	✓	✓	✓	✓	✓	✓
short coil							✓					✓			✓

[key: 1 = standard (helix, strand, coil), 2 = standard + turn, 3 = standard + turn + bend, 4 = standard + exotic (3-turn-helix, 5-turn-helix, β-bridge), 5 = standard + exotic + turn, 6 = standard + exotic + turn + bend, 7 = DSSP + short coil distinction, 8 = (long + short) (helices + strand) + β-bridge + coil, 9 = (long + short) (helices + strand) + β-bridge + coil + turn, 10 = (long + short) (helices + strand) + β-bridge + coil + turn + bend, 11 = (long + short) (helices + strand + coil) + β-bridge + turn + bend, 12 = ((N+C+I)-long + short) (helices + strand) + β-bridge + coil, 13 = ((N+C+I)-long + short) (helices + strand) + β-bridge + coil + turn, 14 = ((N+C+I)-long + short) (helices + strand) + β-bridge + coil + turn + bend, 15 = ((N+C+I)-long + short) (helices + strand) + β-bridge + (long + short) coil + turn + bend; N,C,I mean N-terminal, C-terminal and internal respectively. See also table 4.4. Entries under STRIDE referring to definition levels including bend are missing due to the fact that bend is not resolved by this program.]

Table 4.5: Q_3-accuracy averages in recognition and prediction for different levels of DSSP or STRIDE secondary structure definition (see table 4.4). The STRIDE secondary structure assignment was not tested for levels 7 to 15. The final reduction scheme followed was the loose one defined in section 2.3.1.5. The program's setup was like the one used to obtain the results in table 4.3 with the difference that the scoring functions were optimized using unified statistics. The loose DSSP reduction scheme was applied for both DSSP and STRIDE definitions.

def. level	$|\mathcal{M}|$	$\bar{Q}_3^{\text{(loose)}}$			
		DSSP		STRIDE	
		prediction	recognition	prediction	recognition
1	3	0.6215 ± 0.0007	0.6220 ± 0.0008	0.6172 ± 0.0008	0.6199 ± 0.0007
2	4	0.6016 ± 0.0008	0.6027 ± 0.0009	0.5833 ± 0.0009	0.5849 ± 0.0009
3	5	0.5681 ± 0.0011	0.5706 ± 0.0009	–	–
4	6	0.6221 ± 0.0006	0.6223 ± 0.0008	0.6179 ± 0.0007	0.6206 ± 0.0006
5	7	0.6115 ± 0.0008	0.6121 ± 0.0008	0.5958 ± 0.0008	0.5972 ± 0.0009
6	8	0.5830 ± 0.0010	0.5848 ± 0.0009	–	–
7	9	0.5696 ± 0.0011	0.5751 ± 0.0010	–	–
8	10	0.6185 ± 0.0009	0.6220 ± 0.0008	–	–
9	11	0.6138 ± 0.0008	0.6173 ± 0.0008	–	–
10	12	0.5872 ± 0.0009	0.5903 ± 0.0006	–	–
11	13	0.5463 ± 0.0010	0.5480 ± 0.0009	–	–
12	18	0.5833 ± 0.0007	0.5884 ± 0.0008	–	–
13	19	0.5922 ± 0.0007	0.5973 ± 0.0007	–	–
14	20	0.6077 ± 0.0010	0.6116 ± 0.0007	–	–
15	21	0.6069 ± 0.0007	0.6110 ± 0.0006	–	–

The results indicate that enlarging the number of coil-like classes by separating turns and bends from the bulk meta-class, leads to a loss of accuracy. On the other hand, while it does not bring any concrete improvement, discriminating 3-turn and 5-turn[4] helix types and isolated bridges in the catalogue of secondary structure types did not affect much the performance. Further distinctions based on the length of the SSEs and on the residues' position within the SSEs, turned out to be generally counterproductive. The second remark to be made is that DSSP's secondary structure definition yields generally better sequence-structure correlations than STRIDE's.

On the whole, the idea of increasing the secondary structure definition did not convince because it brought insignificant improvement or none at all. It must be kept in mind, however, that the results were obtained using unified instead of dual statistics. While the heuristic procedure is not applicable to predictors based on too many scoring functions, it is reasonable to believe that precisely then the reweighting of the different secondary structure contributions would be most effective, thanks to the larger number of degrees of freedom. It could be argued that using another technique, like one based on genetic algorithms for instance, it would have been possible to fully exploit the potential of dual

---

[4] The 5-turn helix type can't really be accounted for any sensible alterations of the results due to its negligible statistical weight.

statistics. It is also possible however, to accept this as a limitation of a multi-choice oriented approach and start devising instead a way to avoid such problems rather than overcome them.

4.1.5 Focusing on single-choice classifications

The experiments on secondary structure definition exposed a limitation of the multi-choice oriented approach. As noted before, some fine-tunings like the choice of the reweighting parameters, if rather unorthodox to begin with, become totally unfeasible when scoring functions have to be optimized for a larger number of secondary structure types.

On the other hand, also other characteristics of the classifiers, like the window length or the number of secondary structure key-positions, could be differentiated for each secondary structure of interest. If all these features had to be simultaneously optimized to suit the multi-choice prediction problem directly, the procedure would become too complex because of the huge number of possible combinations of single-choice classifiers.

These arguments eventually induced the focus of the optimization procedures to be shifted towards the individual intermediate single-choice classification problems. The tests to be described in the following were all carried out on the $\mathcal{R}_{90/10}$ subdivisions (see table 2.1 in section 2.3.2.2).

4.1.5.1 Structure-dependent optimization schemes

Of the nine subsets available for learning, only eight were actually learned while one was reserved to perform the so-called supporting test prediction (recall section 2.3.1.1). In order to remove the bias possibly arising from the variable amount of information lost at the N- and C-terminal ends, unit peptides with missing residues were this time allowed (cf. page 47).

The single-choice classification tests require the definition of a threshold likelihood (see section 2.3.2.1) to distinguish between the so-called positive and negative predictions. A threshold likelihood of 0.5 was used here.

Window lengths. To begin with, the choice of the window lengths was reexamined for the different secondary structure types. Peptides sequences with up to 39 residues were tested. The average results obtained in the recognition of the learned data samples and in the prediction of the supporting data samples derived from the $\mathcal{R}_{90/10}$ subdivisions are reported in figures A.1, A.2 and A.3 of appendix A. They clearly show the reduced performance gain attained through use of larger sequence windows. The growth of the quality measures relative to helix and strand is nearly imperceptible when the sequence window reaches a length of 39, while those relative to coil have long since started fluctuating around their ceiling values by that time. No deterioration of the performance like the one experienced in the earliest tests (see figure 4.2) was observed though. It was finally settled to use sequence windows of 19 residues for helix and strand and a sequence window of 17 residues for the coil meta-class, since larger sequence windows offered little or no improvement.

Statistics. Using functions based on the selected sequence windows, the actual effect of dual statistics on the single-choice classification problem was investigated. As the reader may recall from section 4.1.1, non-reweighted dual statistics failed to aid exclusive regression functions (ESFs) in the multi-choice classification. This relatively unexpected behaviour was attributed to the fact that the technique was devised to benefit not the the multi-choice but the single-choice classification.

The helix, strand and coil single-choice classification results for unified and non-reweighted dual statistics are reported, as promised, in table 4.6. The values for specificities and sensitivities, reported here for completeness, immediately jump in the eye for their dissimilarity, but this is probably due to the strong influence that dual statistics exert on the scores and should not be given too much attention. Of more general validity are instead the overall accuracy for positive and negative classification and the Matthews correlation coefficient (third and fourth column respectively). The former seems once again to confirm the thesis that wants unified statistics to perform better than dual statistics (at least with no reweighting). It must be kept in mind however that the overall accuracy index is not very well suited to measure the prediction quality on asymmetric populations of positive and negative samples. In the extreme case in which a population comprised, say, 99% of negatives and only 1% of positives a very simple predictor which always predicts negative would achieve 0.99 accuracy ratio with no effort. Even if to a much smaller extent, the exclusive regression approach inevitably does introduce, as pointed out in section 4.1.1, an asymmetry between positive and negative samples because it classifies as negatives all those samples which do not belong to the target class. The most reliable quality index in this case is therefore the Matthews correlation coefficient (MCC) which is capable of better handling such asymmetries. A brief glance at the fourth columns in table 4.6 confirms then that, as expected, dual statistics perform indeed slightly better in single-choice classification problems.

Reweighting factors. Once the efficacy of dual statistics (even unreweighted) in single-choice classification had been established, the positive/negative reweighting technique was also revised in the new frame. Using scoring functions based on the chosen sequence windows of 19 residues for helix and strand, 17 residues for coil, the performance dependencies on the reweighting factors were analysed. Figure 4.12 shows the quality indices for the recognition of the $\mathcal{R}_{90/10}$ learning subsets plotted against the positive reweighting factor w^+ (see equa-

Table 4.6: Single-choice prediction and recognition performance for helix, strand and coil states, obtained using (a) unified or (b) non-reweighted (reweighting factor $w^+ = 0.5$) dual statistics. The scoring functions used were linear and based on sequence windows comprising 19 residues for helix [H,G,I] and strand [E,B], and 17 residues for coil [C,S,T]. They were optimized to correlate to the secondary structure of the central residue.

(a) unified statistics

	specificity	sensitivity	pos/neg accuracy	MCC	
support prediction	0.682 ± 0.002	0.542 ± 0.002	0.739 ± 0.001	0.420 ± 0.002	helix
	0.708 ± 0.003	0.211 ± 0.003	0.808 ± 0.001	0.312 ± 0.003	strand
	0.713 ± 0.002	0.577 ± 0.002	0.729 ± 0.001	0.431 ± 0.002	coil
recognition	0.683 ± 0.001	0.544 ± 0.002	0.740 ± 0.001	0.420 ± 0.001	helix
	0.709 ± 0.001	0.211 ± 0.001	0.807 ± 0.001	0.312 ± 0.001	strand
	0.714 ± 0.001	0.580 ± 0.001	0.730 ± 0.001	0.434 ± 0.001	coil

(b) non-reweighted dual statistics

	specificity	sensitivity	pos/neg accuracy	MCC	
support prediction	0.584 ± 0.002	0.740 ± 0.002	0.711 ± 0.001	0.420 ± 0.002	helix
	0.413 ± 0.001	0.705 ± 0.002	0.714 ± 0.001	0.361 ± 0.002	strand
	0.663 ± 0.002	0.694 ± 0.002	0.728 ± 0.001	0.443 ± 0.002	coil
recognition	0.582 ± 0.001	0.741 ± 0.001	0.712 ± 0.001	0.421 ± 0.001	helix
	0.415 ± 0.001	0.706 ± 0.001	0.714 ± 0.001	0.362 ± 0.001	strand
	0.665 ± 0.001	0.694 ± 0.001	0.729 ± 0.001	0.445 ± 0.001	coil

tion 4.1.11). The quality indices for the prediction of the supporting data sets resemble very closely the ones for the recognition and are therefore not reported.

Looking at the Matthews correlation coefficients for the reasons explained earlier, the optimal reweighting factors could be located around 0.4 for helix and strand, and around 0.5 for the coil meta-class.

Structure key-positions. Another relevant trait of the single-choice classifiers that was prone to be reassessed within the frame of the single-choice classification problem, was the number of key-positions to be utilized (recall section 4.1.3). More single-choice classification tests were therefore carried out with the intent to establish how many scoring functions for each secondary structure type the multi-choice predictor would count.

Surprisingly, these tests indicated that the use of more key-positions had no relevant effect on the performance. This was probably due to the fact that partially defined peptides, that is peptides with missing residues, had this time been included in the learning data set, and supplied the necessary correlation information for the residues at the N- and C-terminal ends. The earlier registered success in the use of off-center key-positions was in other words to be attributed to the missing information these made up for.

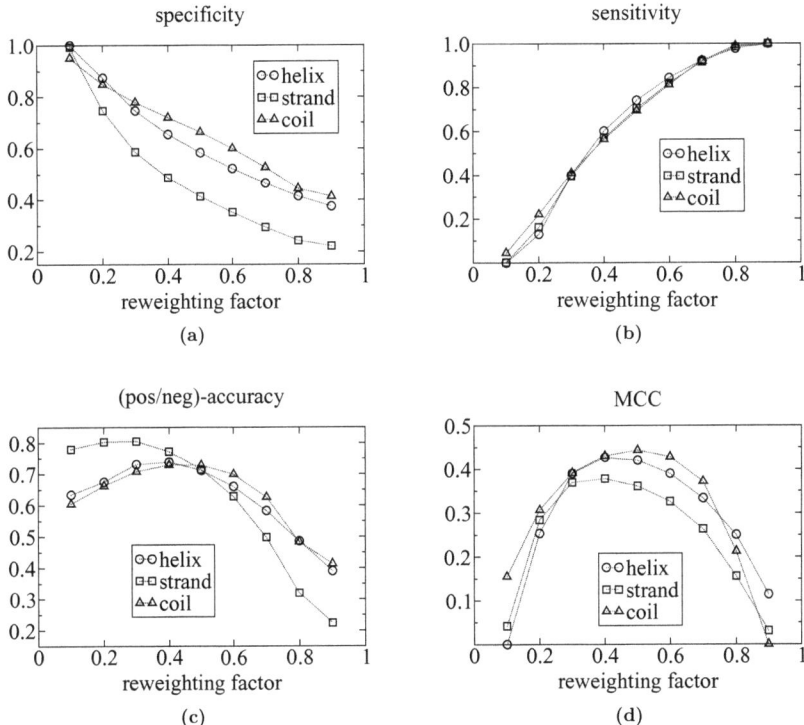

Figure 4.12: Single-choice recognition quality indices (see figure 2.11) dependence on the (positive) reweighting factor (see equation 4.1.11) for helix, strand and coil. The average values and their deviations were computed using equations (2.3.1) and (2.3.2). The deviations were too small to be reported. The functions were based on sequence windows comprising 19 amino acids for helix [H,G,I] and strand [E,B], and 17 amino acids for coil [C,S,T], and were optimized to correlate to the central key-positions only. Quite expectedly, the values of specificity (a) and sensitivity (b) follow opposite behaviours. More interesting from a general point of view are the positive/negative accuracy indices (c) and especially the Matthews correlation coefficients (d).

Table 4.7: Multi-choice prediction and recognition performance for helix, strand and coil states, obtained using reweighted (positive reweighting factors: $w^+ = 0.4$ for helix and strand, $w^+ = 0.5$ for coil) dual statistics. The functions used were linear and based on sequence windows of 19 residues for helix and strand, and of 17 residues for coil. Only the central secondary structure key-positions were employed. The averages and deviations were computed as in equations (2.3.1) and (2.3.2), but for recognition only among a randomly selected subset of roughly the same size as the test set.

		prediction		recognition	
	DSSP	Sens$_3$	Spec$_3$	Sens$_3$	Spec$_3$
helix	[H,G,I]	0.6020 ± 0.0056	0.6577 ± 0.0059	0.6033 ± 0.0049	0.6645 ± 0.0036
strand	[E,B]	0.5419 ± 0.0056	0.5038 ± 0.0094	0.5490 ± 0.0066	0.5083 ± 0.0078
coil	[C,S,T]	0.6880 ± 0.0048	0.6653 ± 0.0050	0.6911 ± 0.0030	0.6651 ± 0.0033
$\bar{Q}_3^{(\text{loose})}$		**0.6243** ± 0.0032		**0.6275** ± 0.0031	

4.1.5.2 Multi-choice classification tests

All was left to do now was to apply the newly optimized single-choice classifiers to the multi-choice classification problem. Even if they are, strictly speaking, not one to one comparable, the results reported on table 4.7, obtained on the ten $\mathcal{R}_{90/10}$ subdivisions, appear to show no sensible improvement with respect to those on table 4.3. The only remarkable effect of the new optimization system regarded the partial accuracy indices. In particular, despite an even less pronounced tendency to predict strand the predictor's sensitivity to the latter increased, at the expenses, apparently, of the one to coil. As far as the overall accuracy is concerned, the recognition Q_3 increased slightly, probably thanks to the larger sequence windows used, but the prediction Q_3 lay still within the statistical fluctuations, and even assuming to be in the presence of a trace of overfit there was nothing to be gained by fighting it.

While these tests did indicate no relevant progress, they at least dismissed the earlier (see page 48) hypothesized intrinsic drawback of employing sequence windows with more than 13 residues. The results of figure 4.2 are then to be interpreted trivially as mere artifacts of the particular way the learning data samples were extracted from the protein domains.

4.2 Quadratic scoring functions

If linear scoring functions are the simplest estimators which can be employed to build a predictor with a reasonable predictive power, quadratic scoring functions are one step ahead in the hierarchy. The general quadratic extensions of the parametrical estimators (4.1.2) are

$$f(W, \mathbf{w}, w_0 ; \mathbf{x}) = \mathbf{x}^t W \mathbf{x} + \mathbf{w} \cdot \mathbf{x} + w_0, \qquad (4.2.1)$$

with $W \in (\mathbb{R}^n \times \mathbb{R}^n)$, $\mathbf{w} \in \mathbb{R}^n$ and $w_0 \in \mathbb{R}$.. Though quadratic in the sequence vector these functions keep being linear in the parameter space. With this in mind, upon introduction of the vectors

$$\tilde{\mathbf{w}} = (w_1, \ldots, w_n, W_{1,1}, \ldots, (2-\delta_{hk})W_{h,k}, \ldots, W_{n,n})$$
$$\tilde{\mathbf{x}} = \left(x_1, \ldots, x_n, x_1^2, \ldots, x_h x_k, \ldots, x_n^2\right)$$
(4.2.2)

where the restriction $h \geq k$ can be imposed thanks to the symmetry of the system, and of

$$\tilde{n} = n + \frac{n(n+1)}{2},$$

the functions (4.2.1) can be rewritten as

$$f(\tilde{\mathbf{w}}, w_0; \tilde{\mathbf{x}}) = \tilde{\mathbf{w}} \cdot \tilde{\mathbf{x}} + w_0, \quad \tilde{\mathbf{w}}, \tilde{\mathbf{x}} \in \mathbb{R}^{\tilde{n}},$$
(4.2.3)

which are formally identical to those in (4.1.2).

This reformulation of the quadratic scoring functions allows to apply the same optimization procedure that was applied in the case of linear scoring functions, so that the task will similarly be that of solving a linear equations system of the kind (recall equation (4.1.12))

$$\left(\tilde{\Gamma}_D^* + \lambda\right)\tilde{\mathbf{w}} = \tilde{\mathbf{b}}_D^*.$$
(4.2.4)

It must be stressed here however that the number of degrees of freedom is much greater in the quadratic case than it is in the linear one. Some examples of quadratic scoring function sizes are collected in the table in figure 4.13a and compared to those of the corresponding linear ones just to give an idea of the orders of magnitude involved.

Effective degrees of freedom. When using the standard sequence representation, as is the case here, it is possible to partially reduce the number of quadratic factors actually entering the functions. Not all the products of sequence vector entries express in fact actual amino acid sequence features and some of them contain just redundant information. As a matter of fact, it is possible to discard all products involving two entries relative to the same amino acid position in the sequence window: if the two factors in the product are relative to the same amino acid type the quadratic term is effectively equivalent to a linear one and therefore redundant; if on the other hand the two factors are relative to two different amino acid types the quadratic term represents a physical impossibility and can therefore be discarded. Assuming an alphabet of 20 amino acids the effective number of parameters entering the quadratic scoring functions, also reported on table 4.13a, is then

$$\tilde{n}_{\text{eff}} = n + \frac{n(n-20)}{2}.$$

	number of equations		
L	$n+1$	$\tilde{n}+1$	$\tilde{n}_{\text{eff}}+1$
5	101	5151	4101
7	141	10011	8541
9	181	16471	14581
11	221	24531	22221
13	261	34191	31461
15	301	45451	42301
17	341	58311	54741
19	381	72771	68781

(a)

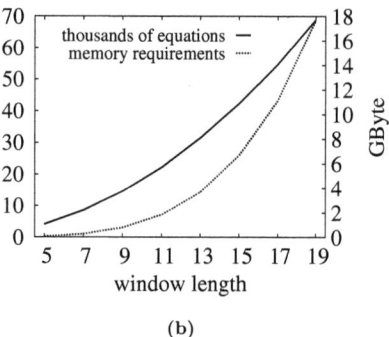

(b)

Figure 4.13: Resource requirements of quadratic scoring functions based on different window lengths. Full amino acid definition is implied here (see section 4.2.2). In the table (a) the number of parameters entering linear and quadratic scoring functions (total $\tilde{n}$, and effective $\tilde{n}_{\text{eff}}$) are compared. The plot (b) displays $\tilde{n}_{\text{eff}}$ and the memory required to store the coefficient matrix of the linear equations system.

Remark: When utilizing profiles instead of standard sequence vectors this reduction could be exclusively applied to the diagonal terms, that is to the terms relative to squared sequence vector entries. It would therefore be practically of no advantage in that case.

The number of degrees of freedom introduced with the quadratic terms requires also a considerable amount of memory especially in order to store the covariance matrix $\tilde{\Gamma}_D^*$ that enters the linear equations system. The memory requirements for quadratic scoring functions based on different window lengths are reported in figure 4.13b together with the number of equations to be solved.

Test results

The quadratic scoring functions were tested in several contexts, but because of their high computational cost not always in a systematical way. At occasions, only the first of the ten $\mathcal{R}_{90/10}$ data set subdivisions was used, with a consequent loss of statistical relevance. At any rate, the number of residues contained in the $\mathcal{R}_{90/10}$ subdivisions ensures a measure of self-averaging, which in turn guarantees quite general validity even for results obtained on a single one of them.

As done for the linear scoring functions, the performances of quadratic scoring functions based on different window lengths were investigated first. In order to limit the computational load while maintaining, to begin with, the full statistical relevance of the results, the sequence windows were limited to contain 5 to 15 amino acids.

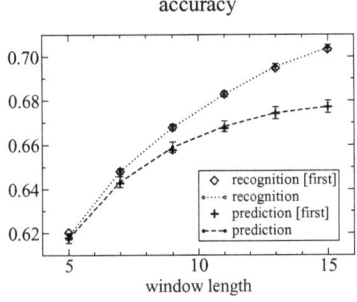

	$\bar{Q}_3^{(\text{loose})}$ $[Q_3^{(\text{loose})}$ – first]	
L	prediction	recognition
5	0.6175 ± 0.0020 [0.6179]	0.6185 ± 0.0014 [0.6202]
7	0.6433 ± 0.0025 [0.6428]	0.6481 ± 0.0012 [0.6480]
9	0.6590 ± 0.0024 [0.6576]	0.6680 ± 0.0013 [0.6678]
11	0.6683 ± 0.0024 [0.6675]	0.6829 ± 0.0013 [0.6829]
13	0.6744 ± 0.0027 [0.6742]	0.6953 ± 0.0014 [0.6946]
15	0.6772 ± 0.0028 [0.6761]	0.7039 ± 0.0015 [0.7043]

Figure 4.14: Q_3-accuracy (loose DSSP reduction scheme) for prediction and recognition, averaged on the $\mathcal{R}_{90/10}$ subdivisions and in detail for the first subdivision, as a function of the window length L. The scoring functions were all optimized using reweighted dual statistics to correlate to the central key-position. The regularization parameter λ amounted to 10^{-5}. The time required to optimize the scoring functions is given in figure 4.15, which reports the time required to assemble the sequence information (i.e. to generate the coefficient matrix of the linear equations system) and to solve the linear equations system. The recognition averages and deviations were computed only for a randomly selected subset of roughly the same size as the test set.

Figure 4.14 shows the average Q_3-accuracy of quadratic scoring functions based on sequence windows of different lengths. It shows a consistent increase in performance with respect to linear scoring functions. While optimistic projections placed the Q_3-accuracy obtainable with the latter somewhere around 0.63 (see section 4.1.5.2), the quadratic scoring functions based on sequence windows of 15 residues reached a Q_3-accuracy of nearly 0.68.

Evident from figure 4.14 is also the widening gap between recognition and prediction, a clear symptom of parameters overfitting. While for linear scoring functions this gap is very small and only mildly increasing with the window length (and therefore the number of parameters), for quadratic scoring functions it grows relatively quickly and amounts to nearly three percent for

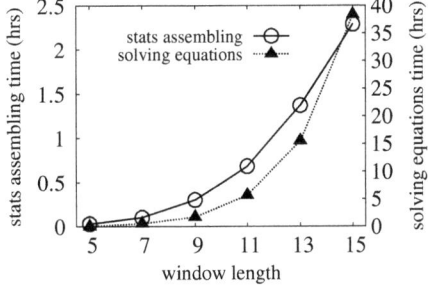

Figure 4.15: Computing time required by a single learning procedure (i.e. to optimize a single scoring function) as a function of the window length. the "stats assembling time" indicates the time required to generate the coefficient matrix of the linear equations system. Full amino acid definition is implied (see section 4.2.2). The computations were performed on an AMD Opteron™275 2.2 GHz processor.

sequence windows comprising 15 residues.

Similar prediction and recognition tests were subsequently carried out on the first $\mathcal{R}_{90/10}$ subdivision using scoring functions respectively based on sequence windows of 17 and 19 residues. While the recognition accuracy maintained an upward trend, with $Q_3 = 0.7136$ and $Q_3 = 0.7210$ for window lengths of 17 and 19 respectively, the prediction accuracy ($Q_3 = 0.6757$ and $Q_3 = 0.6745$ for window lengths of 17 and 19 respectively) was lower than the one achieved on the same subdivision (see table in figure 4.14) using sequence windows with 15 residues. This is an indication that the overfitting had reached such an extent to diminish the predictor's generalization capability.

4.2.1 Reducing overfitting

There are several techniques to prevent overfitting. These can be very different in the way they are applied but all basically consist in *reducing* the number of parameters in the attempt of preventing non-general features of the learned data from being overrepresented.

The most frequently employed techniques are Principal Component Analysis (PCA) [132] and Independent Component Analysis (ICA) [133]. These allow to automatically identify and select a stack of highly significant representatives from the original parameter set, but are computationally quite costly. Alternatively, it is possible to partially *inhibit* the parameters entering the optimization scheme by tuning regularization terms like the one seen in equation (4.1.6). The regularization technique can soothe overfitting by diminishing the *strength* of the parameters, but it does not reduce the parameter space. It therefore falls short of one of the advantages of other techniques, that of lowering the model's resource requirements. And on the other hand the regularization parameter $\lambda = 10^{-5}$ used to obtain the results of figure 4.14 was already such to prevent overfitting to some extent, so that there probably was not much to be gained by further fine-tuning it.

4.2.2 Clustering amino acids

There exists another way to reduce the degrees of freedom of amino acid modelsystems described, as here, by means of the standard sequence representation (see section 3.1.1). This consists in grouping amino acids in *clusters* and use an alphabet of cluster types in place of one of amino acid types. A proper clustering can shrink the dimension of the sequence vector space (see section 3.1) *modifying* as well as reducing the model's parameters.

Ideally, each cluster should contain amino acids that share certain features among each other. Within the project presented here, two different clustering schemes were used. These are shown in figure 4.16.

	ten clusters	
polar		C,Q,N,S,T
nonpolar		A,V,M,L,I
negative charge		E,D
positive charge		K,R
phenylalanine		F
tryptophan		W
tyrosine		Y
histidine		H
glycine		G
proline		P

(a)

	six clusters	
polar		C,H,Q,N,S,T,Y
nonpolar		A,V,M,L,I,F,W
negative charge		E,D
positive charge		K,R
glycine		G
proline		P

(b)

Figure 4.16: Examples of possible amino acids clusterings in (a) ten categories and (b) six categories respectively. Both clustering criteria are based on physicochemical properties of the amino acids, like overall electric charge and hydropathy. Due to their peculiarity though, some amino acids, like, for example glycine and proline, are typically assigned a cluster of their own.

The amino acid clustering can be mathematically translated by replacing the functions (4.2.1) with

$$f(W, \mathbf{w}, w_0 ; \mathbf{x}) = \hat{\mathbf{x}}^t W \hat{\mathbf{x}} + \mathbf{w} \cdot \mathbf{x} + w_0 \qquad (4.2.5)$$

where

$$\hat{\mathbf{x}} = C\mathbf{x}, \qquad (4.2.6)$$

and C is a *clustering matrix*, which projects the standard sequence vector into a vector of lower dimensions. Notice that C is applied only to the quadratic term, being this the most likely source of overfitting for the model. Assuming to be applying, for instance, the clustering of figure 4.16b exclusively to the N- and C-terminal positions of a unit peptide of length L, the clustering matrix C would be the $(L \times L)$-blocks matrix

$$C = \begin{pmatrix} C_6 & 0 & \ldots & & \ldots & 0 \\ 0 & \mathbb{1}_{20} & 0 & & \ldots & \ldots \\ & & \ldots & & & \\ \ldots & \ldots & 0 & \mathbb{1}_{20} & 0 \\ 0 & \ldots & & \ldots & 0 & C_6 \end{pmatrix},$$

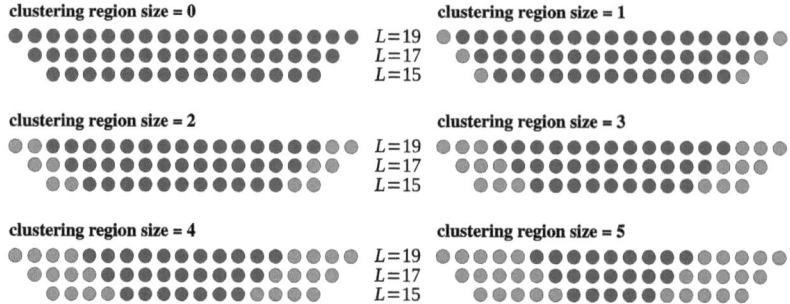

Figure 4.17: Illustration of amino acid clustering with different clustering zones sizes. In blue the amino acid positions at which the clustering was not applied, in orange the amino acid positions at which the clustering was applied.

where

$$C_6 = \begin{pmatrix} 0 & 0 & 1 & 0 & 1 & 0 & 1 & 0 & 1 & 0 & 0 & 0 & 0 & 0 & 1 & 1 & 0 & 1 & 0 \\ 1 & 0 & 0 & 0 & 0 & 0 & 0 & 0 & 1 & 1 & 0 & 1 & 1 & 0 & 0 & 0 & 1 & 0 & 1 \\ 0 & 0 & 0 & 1 & 0 & 1 & 0 & 0 & 0 & 0 & 0 & 0 & 0 & 0 & 0 & 0 & 0 & 0 & 0 \\ 0 & 1 & 0 & 0 & 0 & 0 & 0 & 0 & 0 & 0 & 1 & 0 & 0 & 0 & 0 & 0 & 0 & 0 & 0 \\ 0 & 0 & 0 & 0 & 0 & 0 & 1 & 0 & 0 & 0 & 0 & 0 & 0 & 0 & 0 & 0 & 0 & 0 & 0 \\ 0 & 0 & 0 & 0 & 0 & 0 & 0 & 0 & 0 & 0 & 0 & 0 & 0 & 1 & 0 & 0 & 0 & 0 & 0 \end{pmatrix} \text{6 rows}$$

20 columns

and alphabetical ordering of the amino acids is assumed, as in figure 1.1 of the introduction.

The amino acid clustering was applied to quadratic scoring functions based on sequence windows of length 15, 17 and 19, the ones which showed strongest overfitting. Only the N- and C-terminal regions were interested by the clustering. Full amino acid definition was preserved for the inner positions. Figure 4.17 illustrates the different clustering patterns applied. The dimensions of the parameter spaces resulting from the different clustering patters are collected in table 4.8. The results of the test prediction performed on the first $\mathcal{R}_{90/10}$ subdivision are reported in figure 4.18.

While the clustering did reduce the recognition quality (the results are not shown), it had, though mildly beneficial, a far less relevant influence on the prediction quality. The best Q_3-accuracy index of about 0.682 was obtained using sequence windows of 19 amino acids and clustering in six categories at four+four terminal positions.

Table 4.8: Number of parameters in quadratic scoring functions with different clustering zone sizes for two different types of clustering.

clustering regions size	ten clusters			six clusters		
	$L = 15$	$L = 17$	$L = 19$	$L = 15$	$L = 17$	$L = 19$
0 (full)	45451	58311	72771	45451	58311	72771
1	39641	51701	65361	37429	49169	62509
2	34231	45491	58351	30191	40811	53031
3	29221	39681	51741	23737	33237	44337
4	24611	34271	45531	18067	26447	36427
5	20401	29261	39721	13181	20441	29301

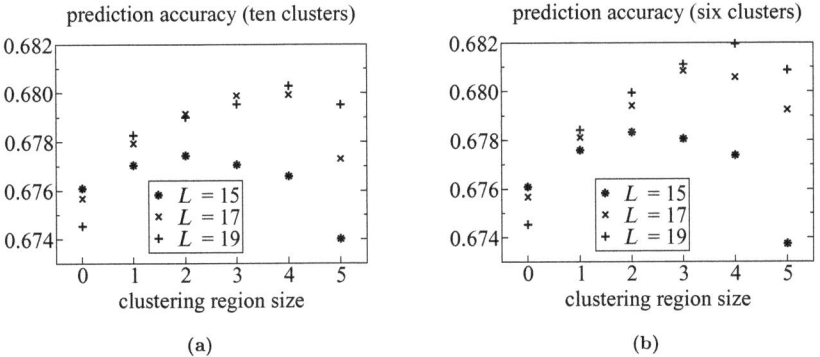

Figure 4.18: Q_3-accuracy (loose reduction scheme) in prediction using partial amino acid clustering, respectively in (a) ten and (b) six categories. The abscissa shows the number of positions at the window's N and C termini at which amino acids were actually clustered, zero corresponding to full definition (see figure 4.16a). The scoring functions were based on sequence windows of 15, 17 and 19 amino acids.

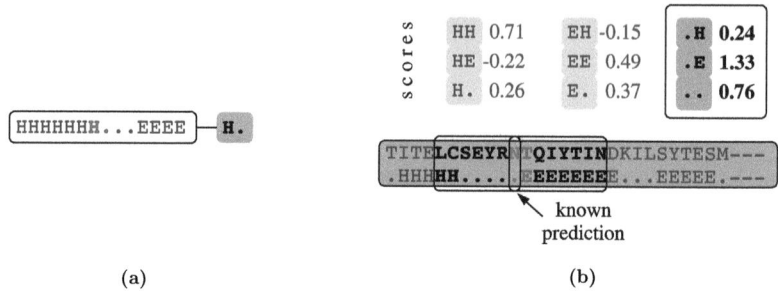

Figure 4.19: Usage of extended secondary structure patterns. In (a) a two-dimensional secondary structure pattern assignment with sequence window 14 (even window lengths are preferably used when the secondary structure patterns comprise an even number of amino acids) and key-position 7. In the learning phase the unit peptide in this example is assigned the pattern "H.". In (b) a possible prediction scenario in which scoring functions optimized to discern two-dimensional secondary structure patterns are employed: in predicting the secondary structure motif of the threonine residue (in cerise) six of the original nine scores can be ignored following the indication of a previous prediction (in green).

4.2.3 Extended secondary structure patterns

One concept developed in the course of the project, which eventually turned out to be quite unsuccessful, but nevertheless deserves to be mentioned, is that of extended secondary structure patterns. Some examples of secondary structure patterns involving more than one residue already made their appearance when the concept of secondary structure library was introduced in section 2.3.1.4.

The rationale for devising this construct was not dissimilar from the one behind the investigations on secondary structure definition analysed in section 4.1.4 of this report: considering extended patterns should yield a sharper secondary structure definition.

The learning procedure carried out to optimize the scoring functions for extended secondary structure patterns is in all respects identical to the one carried out for the scoring functions analysed so far. As usual for exclusive regression functions (ESF), the secondary structure pattern has to be read at a defined key-position. The extended secondary structure patterns most widely employed in this work were the simplest ones formed by two consecutive secondary structure motifs. In applications involving such secondary structure patterns, the two motifs starting from the one at the key-position itself are used to build the two-dimensional secondary structure pattern, as shown in figure 4.19a.

The most delicate issue faced while employing scoring functions optimized to correlate to extended secondary structure patterns, concerned rather how to actually fit them into the prediction scheme. The goal remained in fact that of predicting the secondary structure motif for every *single* residue in a protein chain. For such a task, secondary structure patterns involving *more than* one residue seem quite ill suited. A method had to be devised, which would enable to use the likelihood of patterns of arbitrary size in a residue-wise secondary structure prediction. The following example, illustrated also in figure 4.19b will hopefully give a better insight into it.

Example: Let's take the simplest extended patterns, those composed by the secondary structure motifs of two consecutive amino acids, and assume a set of scoring functions has been optimized to distinguish them. Let's assume also that a prediction is available for an amino acid in a certain protein chain and that a prediction for the one right next to it is sought. To begin with, the scoring functions for the extended secondary structure patterns are evaluated on the amino acid sequence pertaining to the designated motifs. This gives nine scores. Recalling then however the secondary structure information from the available prediction, the spectrum of possible choices can be collapsed down to three and the problem's complexity is thus reduced back to the usual one. The secondary structure prediction of the second amino acid can be then utilized in that of a third and so on.

The same technique just exemplified can of course be applied to extended patterns involving any number of amino acids. Independently from the number of amino acids however, the crucial feature of the whole procedure, as the attentive reader may already have guessed, is that it must be *initialized*.

The first kind of initialization applied was a very trivial one: the prediction started at the N terminus of every protein chain, assuming the initial secondary structure motif to be coil. This lead however to very poor results. Even using quadratic scoring functions based on window lengths of 14 amino acids and optimized with unified statistics, the Q_3 prediction accuracy achieved on the first ASTRAL40-based $\mathcal{R}_{90/10}$ subdivision was barely of about 0.52. Clearly enough it was necessary to improve the initialization scheme.

4.2.3.1 Initialized prediction

The basis of the new initialization concept was the idea of having not one but possibly several initialized secondary structure *seeds* for every peptide chain, as depicted in figure 4.20. These would serve as multiple starting points for the prediction which would then proceed as exemplified in figure 4.19 in both directions along the chain.

Using an artificial initialization grid with a minimum seed-seed distance of eight residues, various scenarios were simulated in which the residues in the grid were assigned with increasing probability the correct secondary structure motif, or with decreasing probability a randomly chosen one. The scoring functions

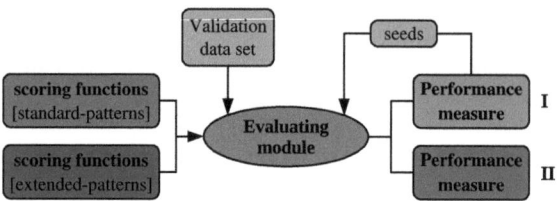

Figure 4.21: Scheme of the two-phase prediction procedure. Phase I involves scoring functions optimized to recognize standard single-residue secondary structure patterns and serves to identify the potential secondary structure seeds. Phase II involves scoring functions optimized to recognize extended secondary structure patterns. It uses the seeds identified in phase I.

employed to measure the likelihood of the two-dimensional secondary structure patterns were based on a sequence window comprising 14 amino acids. They were optimized using unified statistics. The regularization parameter was maintained constant at $\lambda = 10^{-5}$. The results of the prediction simulation are reported in figure 4.22a.

The plot shows a nearly linear correlation between the seed accuracy and the prediction accuracy. It is interesting to observe that even very accurate seeds lead only to a moderately good prediction performance. Depending on the density and quality of the actual seeds however, the chance that some improvement might still come about could not be ruled out.

Figure 4.20: Illustration of an ideal seed-based initialized prediction scenario: the procedure starts from multiple initialization seeds. Note that the initialization seeds are all within helix or strand SSEs.

The only question that remains unanswered at this point is how the initialization points could actually be designated. A pre-prediction phase (phase I), as illustrated in figure 4.21, would serve exactly to this purpose. This phase would involve traditional scoring functions like the ones used to obtain the results of figure 4.14, optimized to recognize single-residue secondary structure patterns. The *safe* residues identified during the pre-prediction would then be used as initialization points in the actual prediction phase (phase II), which would instead involve extended secondary structure patterns. Key node of the procedure was the identification of the safe residues. These should in fact provide reasonably secure starting points, but also guarantee a relatively dense initialization grid. This task was accomplished by analysing the prediction scores from the first phase and by making use of two old acquaintances from section 4.1.2.1, the safety and criticality thresholds t_s and t_c, again normalized with respect to each other. The scheme prescribed a residue to be safe, provided the highest

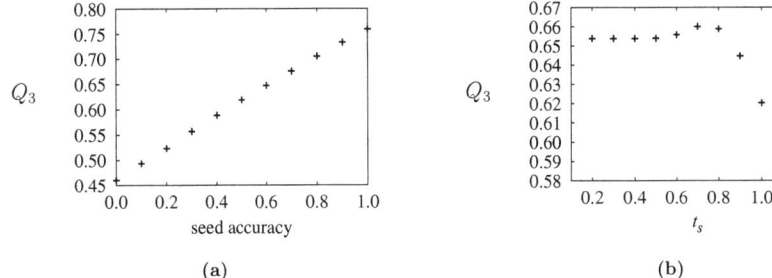

Figure 4.22: Extended secondary structure patterns. In (a) the Q_3-accuracy in a seed-based secondary structure prediction simulation. The seed accuracy at the abscissa represents the ratio of seeds being assigned the correct secondary structure motifs. The remaining were assigned random motifs. Scoring functions based on sequence windows of 14 amino acids were used for all extended secondary structure patterns. In (b) the Q_3-accuracy as a function of the seed safety threshold t_s for a real seed-based secondary structure prediction.

score it achieved was above t_s and the difference between the two highest scores was above $(t_s - t_c)$. Safe residues were also restricted to belong to either helix or strand SSEs, since starting the prediction in the middle of a coil region was reckoned to be counterproductive.

The scoring functions relative to the standard secondary structure patterns were based on sequence windows of 15 amino acids, the ones relative to the extended ones were based on sequence windows of 14 amino acids. The latter were again optimized using unified statistics.

The threshold-dependent prediction performance is reported in figure 4.22b. The plot shows an initial plateau followed by a modest rise and a downfall. The plateau corresponds probably to the initial indifference in the choice of very unrestrictive thresholds. The subsequent increase in thresholds demands apparently improves the overall quality of the secondary structure seeds and as a consequence that of the prediction. When the thresholds become too strict however, the number of residues meeting their requirements drops so much that the seeds get too sparse to constitute an efficient initialization grid. Even the best prediction performance however, achieved using $t_s = 0.7$, lay around 0.66, that is more than two percent lower than that achieved with standard quadratic scoring functions. All in all therefore, the test outcome did not speak in favour of the new technique.

The perfect grid. It is of some interest to see what would become of the prediction performance if a *perfect* grid was used. In order to show this, the procedure described above was applied making use of an all-residue dense grid

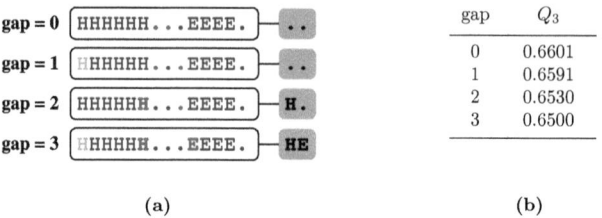

Figure 4.23: Gapped extended secondary structure patterns. In (a) is illustrated how different patterns can be assigned to the same amino acid sequence, eventually parity-adjusted to the effective size of the pattern itself. In (b) the effects of the use of different gaps on the prediction performance.

with 100% percent seed accuracy, while the rest of the settings where left exactly like in the previous tests. Interestingly, the Q_3 accuracy performance increased then up to roughly 0.87, very close to the hypothesized theoretical limit [121].

4.2.3.2 Gapped secondary structure patterns

An attempt to rescue the idea of extended secondary structure patterns was made by modifying the way these were defined in the first place. This was done, as illustrated in figure 4.23a by introducing variable gaps between the amino acids actually entering a secondary structure pattern. To encourage this final attempt was the idea that two adjacent secondary structure motifs might be too closely related to represent genuine persistence or transition sequences. Unfortunately, increasing the gap caused instead a drop in performance as can be read from table 4.23b.

Chapter 5
Profile-based predictor

The present chapter is dedicated to the analysis of the most crucial step in the development of the project: the evolution from standard sequence-based to profile-based scoring functions.

PSI-BLAST profiles

The sequence profiles employed to instruct the new predictor are a by-product of the multiple sequence alignment tool PSI-BLAST [134]. They emerged (see figure 5.1) from the third PSI-BLAST iteration in the alignment of 4,396,331 protein sequences belonging to a non-redundant database including the GenPept [135], Swiss-Prot [136], PDB [98], PIR [137], PRF [138], and NCBI RefSeq [139] databases. An example of a PSI-BLAST profile is reported in figure 5.2.

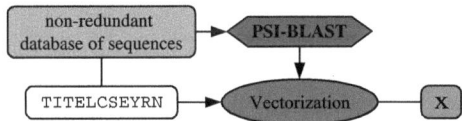

Figure 5.1: Scheme of the profile generation process. While the generation of a standard sequence vector involves only the sequence of interest, as shown in figure 3.1, that of a profile actually takes a whole database of sequences.

5.1 Preliminary tests: a quality leap

The enormous potential inherent to the profile representation of amino acid sequences was made clear very soon, in a preliminary test carried out on the first of the $\mathcal{R}_{90/10}$ subdivisions. Using quadratic scoring functions based on a sequence window comprising eleven amino acids and optimized with reweighted dual statistics, it was possible to achieve a prediction Q_3-accuracy of 0.784. Not only the

```
#TDB 9wgaA0   171 WHEAT GERM AGGLUTININ (ISOLECTIN 2)
-    0 -1 -1 -1 -2 -1 -1 -1 -1 -1 -1 -1 -1 -2  0  0 -2 -1 -1
R   -4  6 -3 -4 -6  6 -2 -5 -3 -6 -5  3 -3 -6 -4 -3 -4 -6 -4 -5
C   -3 -5 -6 -6 11 -6 -7 -6 -6 -4 -3 -5 -4 -4 -6 -4 -3 -5 -4 -3
G    0 -2  1  2  0  0  0  3  1 -3 -2 -1 -3 -5  2  0  0 -2 -4 -4
E   -1  0  2  0  6  0  1 -2 -2 -3 -4 -1 -3 -3  3  0  0 -1 -5 -2
Q   -1 -1  2  0  5  1  0  3  0 -3 -4 -2 -3 -4 -2  0  0 -5 -2 -3
G    0 -1  0 -1  7 -1  0  1  0 -2 -3  0 -1  0  0  0  0  3  0 -3
S    0 -1  0  1  3  0  1  1  2 -4 -3 -1 -4 -2 -1  2  0 -1  0 -3
N   -1  0  2  0  3  1  1  3 -2 -4 -5 -2 -3 -3 -1  0  0 -5 -1 -2
...                                     ...
```

Figure 5.2: Example of a PSI-BLAST profile. The entries are integers. Each column refers to a specific amino acid type and represents its affinity to the one in the actual sequence: the larger the number the higher the affinity. Additional information output by PSI-BLAST is not reported here.

threshold of 0.7 had been breached, but even that of 0.8 was all of a sudden within grasp.

5.1.1 Choosing a reduction scheme

All along the development analysed so far, the eight DSSP secondary structure classes were reduced to the three standard ones according to the loose reduction scheme illustrated in section 2.3.1.5. As observed by others [51] before, however, the strict reduction scheme seems to lead to better Q_3-accuracies. In a second test, which in all other respects was identical to the one just mentioned, the prediction Q_3-accuracy achieved using the strict reduction scheme was 0.798, that is more than a point percent better than the one achieved using the loose reduction scheme.

One might wonder which factor exerts a stronger influence on these results: one might ask in particular whether the strict reduction scheme leads to better performance because it prevents the contamination of the helix and strand classes, or simply because it augments the population of the coil class, increasing the disproportion among the classes and thus facilitating the whole secondary structure prediction task. In order to address this question a new predictor was instructed using the strict reduction scheme, but with a slightly modified learning data set. A fraction of the unit peptides in the original data set was discarded so that the proportions of the secondary structure populations resembled those attainable by applying the loose reduction scheme. In particular, only roughly 90% of the coil sequence samples were included in the learning data set. The new predictor achieved a Q_3 of 0.797, very close to the one achieved by the predictor instructed with the complete data set. This indicates that the way the strict reduction scheme affects the proportions of the secondary structure populations does not influence the performance. Hence, the main effect must be the contamination of the helix and strand classes.

In light of these considerations the strict reduction scheme was used in place of the loose one in the following development. While weighing the implications of this decision it became clear that by increasing the sequence-structure correlation potential, the strict reduction scheme could also amplify the effects of upgrades yet to befell to the model and thus be of more practical use as well.

The applications to be discussed in part III of the report attempted to compare the efficacy of the predictor developed to that of a selection of existing ones. Since the latter adopt, for the most part, the loose DSSP reduction scheme, the results presented there will stick back to it.

5.2 Super scoring functions

An important construct, which was introduced concurrently with the sequence profiles and eventually became an integral part of the profile-based predictor, is the *super scoring function*.

> **Historical digression**: The idea of super scoring function originated in the search for a mechanism to synergically combine a number of scoring functions into a single single-state classifier (recall the discussion of multiple secondary structure key-positions in section 4.1.3). Eventually, the concept of using multiple key-positions to build a classifier turned out to be rather unproductive (see end of section 4.1.5.1) and the need to combine them trailed off together with it. It would be the introduction later on of a new type of scoring function to reinvigorate the idea. The new type of scoring function was designed to distinguish between *two* secondary structure patterns only (e.g. distinguish helix from strand), rather than between one and *all* others. The only way the new scores could actually be used however, was to have their contributions automatically weighed together with those of the standard scores. Later still, the super scores further developed into combinations of *all* available scores.

Let's assume a scoring function (which will be also called *bare* scoring function from now on, to distinguish it from the *super* scoring function yet to be defined) $f_\delta \in \mathcal{F}(S, \mathbb{R})$ has been optimized for each single-choice classification problem $\delta \in \mathcal{C}_\mathcal{P}$, that is for both "one versus one" (e.g. helix vs. strand, helix vs. coil, strand vs. coil) and "one versus all" (e.g. helix vs. other, strand vs. other, coil vs. other) classification problems[1].

The problem is now that of finding a means of combining these bare scoring functions into a new *super* scoring function that makes the best possible use of each contribution. The idea is to build this new function by simply taking (in first approximation) a linear combination of the f_δs,

$$F = \sum_{\delta \in \mathcal{C}_\mathcal{P}} u_\delta f_\delta + u_0, \tag{5.2.1}$$

---

[1]If $N_\mathcal{P}$ is the number of secondary structure patterns, the cardinality of $\mathcal{C}_\mathcal{P}$ is $|\mathcal{C}_\mathcal{P}| = N_\mathcal{P} + \binom{N_\mathcal{P}}{2}$.

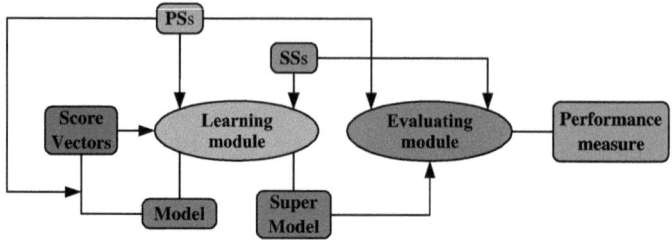

Figure 5.3: Scheme of the program including the super scoring function optimization step. The models provided by the bare scoring functions are used to compute a set of score vectors. These are in turn re-input by the learning module, which constructs the so-called super models. These are then passed on to the evaluation module.

where the parameters $u_\delta \in \mathbb{R}$, $\delta \in \mathcal{C}_\mathcal{P}$, must be optimized, like the parameters of the bare scoring functions, to fit a given data set. Introducing the vectors

$$\mathbf{f} = \left(f_1, \ldots, f_{|\mathcal{C}_\mathcal{P}|}\right)^t \in \mathcal{F}(S, \mathbb{R}^{|\mathcal{C}_\mathcal{P}|}) \quad \text{and} \quad \mathbf{u} = \left(u_1, \ldots, u_{|\mathcal{C}_\mathcal{P}|}\right)^t \in \mathbb{R}^{|\mathcal{C}_\mathcal{P}|},$$

the function (5.2.1) can be rewritten as

$$F = \mathbf{u} \cdot \mathbf{f} + u_0, \qquad (5.2.2)$$

which is formally identical to (4.1.2). The main difference in fact is that instead of a function of the sequence vector, (5.2.2) is a function $F \in \mathcal{F}\left(\mathbb{R}^{|\mathcal{C}_\mathcal{P}|}, \mathbb{R}\right)$ of the *score* vector, that is, the vector whose components are the scores obtained by applying the f_δs to the sequence vector [2]. The super scoring functions were therefore optimized in the same way the bare scoring functions were using the same machinery developed in that case (see figure 5.3).

As for the bare scoring functions-based predictor, the multi-choice classification problem would be addressed employing one super scoring function F_σ for every secondary structure of interest $\sigma \in \mathcal{P}$. The classification procedure would therefore be in all respects identical to the one observed this far.

5.2.1 The score neighbourhood

All scores used by the super scoring functions, independently from the classification problem they stem from, result from the correlation of the sequence to the

---

[2]If needed, an arbitrary number p_δ of scores can still be introduced for each classification problem $\delta \in \mathcal{C}_\mathcal{P}$. The super scoring function becomes in this case

$$F = \sum_{\delta \in \mathcal{C}_\mathcal{P}} \sum_{c=1}^{p_\delta} u_\delta^{(c)} f_\delta^{(c)} + u_0,$$

which is analogous to (5.2.1) and can therefore undergo its same treatment.

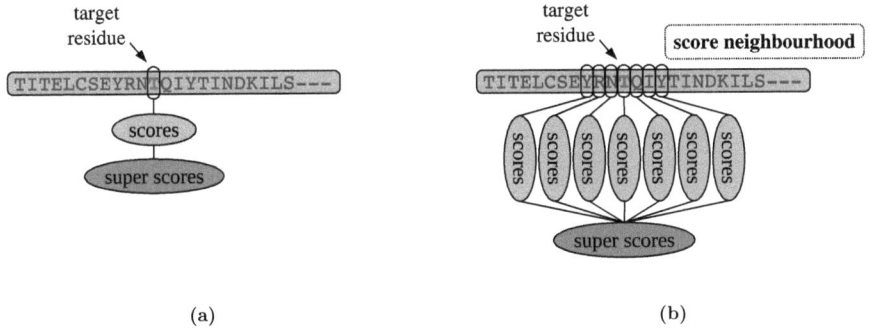

Figure 5.4: Illustration of the evolution from the super scoring function of type (5.2.1) to the new neighbourhood-based (5.2.3). In (a) a set of bare scores are calculated around the target residue and used to generate a super score; in (b) the bare scores are computed for every residue in the neighbourhood $\mathcal{N}$. All sets of scores obtained this way enter the super scoring functions of new design.

secondary structure of a *single* residue. Eventually, a way was found to further extend the super scoring functions correlation reach.

The inspiration really came by attentively observing how the learning procedure was designed in other secondary structure prediction programs. After establishing the correlation rules between the primary and the secondary structure (*sequence-structure* correlation), many of the existing programs also compute correlation rules between the secondary structure of the target residue and that of a number of neighbouring ones (*structure-structure* correlation). This is done, in practice, by applying the learned sequence-structure correlation rules to all the residues in the designated neighbourhood and subsequently establishing the correlation rules between the *scores* obtained in this way, and the secondary structure of the target residue. A scheme depicting the implementation of this technique within the software presented here is reported in figure 5.4b.

In formal terms this translates into upgrading (5.2.1) to

$$F = \sum_{r=1}^{|\mathcal{N}|} \sum_{\delta \in \mathcal{C}_p} u_{r,\delta} f_{r,\delta} + u_0, \qquad (5.2.3)$$

where r labels the residue's address in the neighbourhood set $\mathcal{N}$.

The sequence-structure/structure-structure terminology will be borrowed in the following too to refer to the two levels of scoring function optimization.

5.2.2 Quadratic super scoring functions

The process of enhancement of the super scoring functions did not end with the introduction of the score neighbourhood. As seen in section 4.2, the optimization machinery developed was capable to record second order correlations too and there was no reason why the super scoring functions should be limited instead to the first order.

The general shape of a quadratic super scoring function is

$$F = \sum_{H>K} U_{HK} f_H f_K + \sum_K u_K f_K + u_0, \qquad (5.2.4)$$

where the new indices $H, K = 1, \ldots, n_s$, label, for simplicity, all possible score contributions, that is the ones coming from functions inherent to different single-choice classification problems and different residues of the score neighbourhood.

In complete resemblance to what was done in section 4.2 the functions (5.2.4) can be rewritten as

$$F = \tilde{\mathbf{u}} \cdot \tilde{\mathbf{f}} + u_0, \qquad (5.2.5)$$

upon prior definition, of course, of the vectors

$$\tilde{\mathbf{u}} = (u_1, \ldots, u_{n_s}, U_{2,1}, \ldots, U_{H,K}, \ldots, U_{n_s,n_s-1})$$

and

$$\tilde{\mathbf{f}} = (f_1, \ldots, f_{n_s}, f_2 f_1, \ldots, f_H f_K, \ldots, f_{n_s} f_{n_s-1}).$$

In the form (5.2.5) the true nature of the super scoring function F is again revealed.

5.2.3 Further upgrades

Simultaneously to the introduction of the enhanced super scoring functions, two more aspects of the predictor underwent some upgrades.

Multi-fold statistics. The reweighting was modified, at sequence-structure as well as at structure-structure correlation levels, by introducing *in general* different factors for samples to be assigned different secondary structure patterns. The objective function (structure-structure correlation level) would then be

$$E_M = \sum_{\sigma \in \mathcal{P}} \frac{w_\sigma}{2N_\sigma} \sum_{i=1}^{N_\sigma} \left(F\left(\mathbf{x}_i\right) - y_i \right)^2, \qquad (5.2.6)$$

its minimization resulting in the linear equation system

$$\tilde{\Gamma}_M \tilde{\mathbf{u}} = \tilde{\mathbf{b}}_M, \qquad (5.2.7)$$

where

$$\left[\tilde{\Gamma}_M\right]_{kh} = \left\langle \tilde{f}_k \tilde{f}_h \right\rangle_M - \left\langle \tilde{f}_k \right\rangle_M \left\langle \tilde{f}_h \right\rangle_M$$

$$\left[\tilde{b}_M\right]_k = \left\langle \tilde{f}_k \right\rangle_M \langle y \rangle_M - \left\langle \tilde{f}_k y \right\rangle_M,$$

and, in general,

$$\langle q \rangle_M = \sum_{\sigma \in \mathcal{P}} \frac{w_\sigma}{N_\sigma} \sum_{i=1}^{N_\sigma} q_i. \quad (5.2.8)$$

The sums contained in (5.2.8) are understood as partial sums over subsets pertaining to a single secondary structure class.

The w_σs for helix, strand and coil were optimized independently, with the same technique employed to optimize the reweighting factors in the past. The most convenient choices turned out to be very close to 0.5 for both patterns involved in the "one versus one" functions types, and close to 0.4 and 0.3 for the pattern of interest and for the other two respectively, in the "one versus all" function types.

Improved regularization. The regularization term introduced in equation (4.1.5) affects all scoring function parameters in the same measure. During the development of the profile-based predictor it was improved in that it was made sensitive to the type of parameter it affected. In mathematical terms the equations (4.2.4) were turned into

$$\left(\tilde{\Gamma}_M + \Lambda \right) \tilde{\mathbf{w}} = \tilde{\mathbf{b}}_M, \quad (5.2.9)$$

where Λ is a diagonal *regularization matrix*. Of course it is feasible to differentiate the value of λ up to a certain level only. To begin with, this was done for linear and quadratic terms, setting in other words Λ to a (2×2)-block matrix of the kind

$$\Lambda_2 = \begin{pmatrix} \lambda_{\text{lin}} \mathbf{1}_n & 0 \\ 0 & \lambda_{\text{quad}} \mathbf{1}_{\tilde{n}-n} \end{pmatrix}, \quad (5.2.10)$$

where λ_{lin} and λ_{quad} are the regularization parameters for linear and quadratic terms respectively. Subsequently, the regularization parameters were differentiated for the latter according to the distance, in number of residues, between the two residues correlated. The larger this distance the larger the parameter should be in order to inhibit unlikely correlations. The resulting form of the regularization matrix was then

$$\check{\Lambda}_2 = \begin{pmatrix} \lambda_{\text{lin}} \mathbf{1}_n & 0 \\ 0 & \check{\Lambda}_{\text{quad}} \end{pmatrix}, \quad (5.2.11)$$

where $\check{\Lambda}_{\text{quad}}$ is a diagonal matrix of $\left(\mathbb{R}^{(\tilde{n}-n)} \times \mathbb{R}^{(\tilde{n}-n)} \right)$ the entries of which are proportional to the strength of the inhibition. Tests aimed at identifying the best regularization parameters suggested to apply a very mild distance dependence or none at all. It appeared more convenient in other words to inhibit all quadratic terms in the same way, as when using (5.2.10), rather than favour short range correlations at the expenses of long range ones.

5.2.4 Test results

The bare scoring functions chosen for the final profile-based predictor tests were quadratic and based on sequence windows of 15 amino acids. No filters were applied to the learning data sets. Neighbourhood windows comprising as well 15 residues (for a total of 4186(+1) super parameters) were used to build the super scoring functions.

5.2.4.1 Optimizing the regularization parameters

In order to loose as little performance as possible to overfitting the regularization parameters for the quadratic terms in scoring and super scoring functions were optimized ahead of the multi-choice test itself. This was achieved by monitoring the λ_{quad}-dependence of helix, strand and coil single-choice classification prediction performances. The optimization was carried out on the first of the $\mathcal{R}_{90/10}$ subdivisions, which acted this way as a supporting test data set. For this reason the results obtained on the latter were not taken into account when computing the averages (see section 5.2.4.2) later on.

The results of the λ_{quad} optimization for the bare scoring functions are reported in figures 5.5, 5.6 and 5.7. As can be seen, all quality indices follow very close trends. The initial plateaus mark the region in which λ_{quad} is too small to sensibly affect the scoring function parameters. The subsequent more or less pronounced rise in prediction, accompanied by the initial decline in recognition, indicates the sought after overfit-inhibiting effect of the regularization term. The rise in prediction is followed by a collapse of both prediction and recognition accuracies, most probably caused by "overinhibition". In other words, when the regularization parameters get too large they start to suppress the parameters too strongly thus causing a loss of valuable correlation information.

It is instructive to look at how the parameter vectors $\mathbf{w}$ resulting from the use of different λ_{quad} values actually deviate from each other. The normalized scalar product of two parameter vectors

$$s_{ab} = \frac{\mathbf{w}_a \cdot \mathbf{w}_b}{\|\mathbf{w}_a\| \, \|\mathbf{w}_b\|} \qquad (5.2.12)$$

was used to measure the similarity between them.

Figure 5.8 shows the similarity distribution over the ten $\mathcal{R}_{90/10}$ subdivisions[3], for three representative values of λ_{quad}. The reference similarity distribution was obtained setting $\lambda_{\text{quad}} = 10^{-12}$, a value that causes virtually no effect on the scoring function parameters. Several orders of magnitude above, at the optimal value of the regularization parameter $\lambda_{\text{quad}} = 10^{-4}$, the similarity factors appear to be slightly larger, but their spread is unchanged. It is when the regularization factor is set to 0.9 that its influence is heavily felt. In this regime the parameters

---

[3] for a total of $\binom{10}{2} = 45$ similarity values

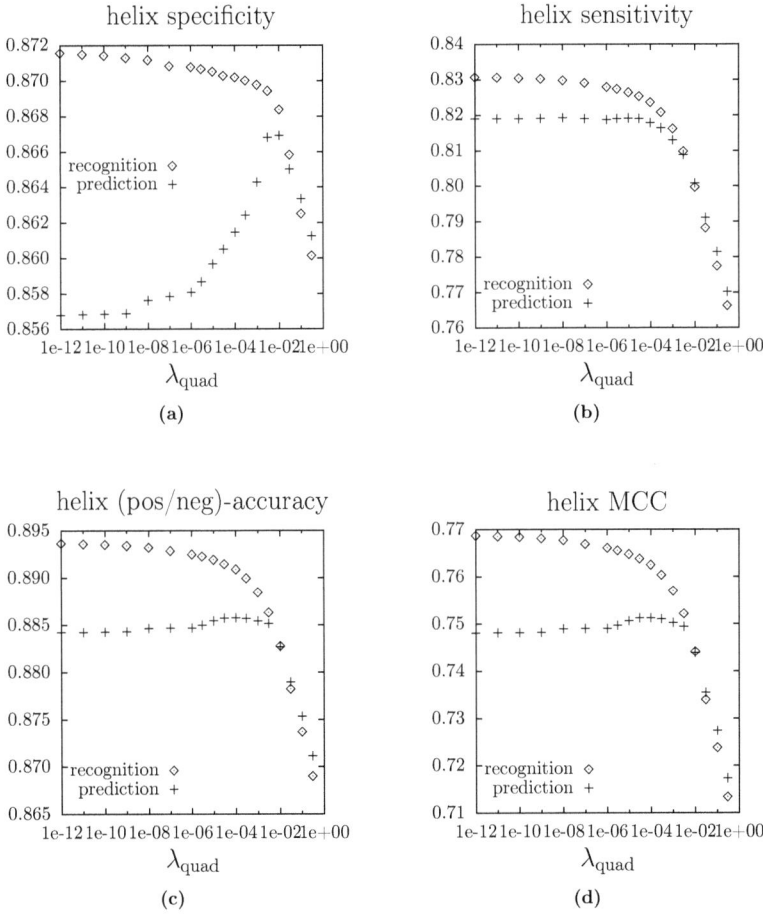

Figure 5.5: Helix [H] recognition and prediction quality indices dependence on the quadratic regularization factor λ_{quad}.

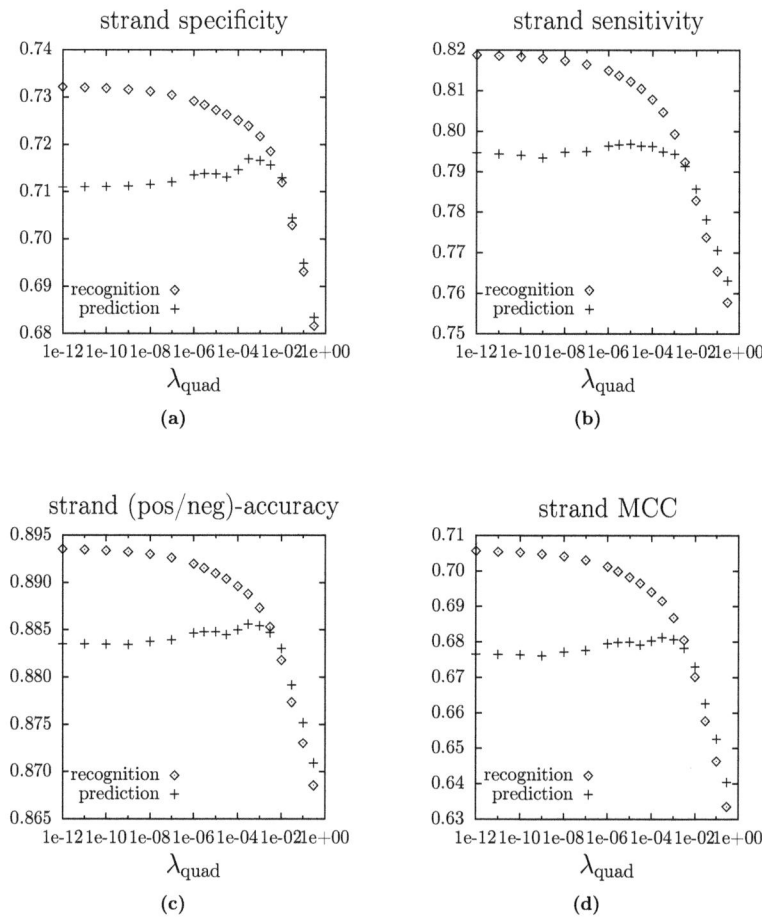

Figure 5.6: Strand [E] recognition and prediction quality indices dependence on the quadratic regularization factor λ_{quad}.

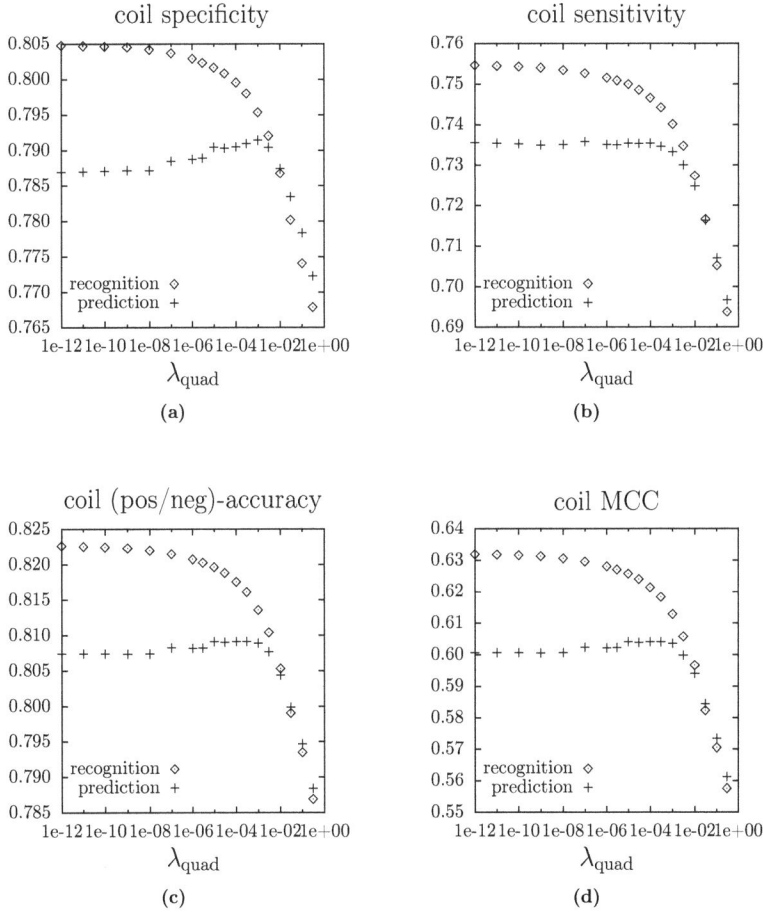

Figure 5.7: Coil [C,S,T,G,I,B] recognition and prediction quality indices dependence on the quadratic regularization factor λ_{quad}.

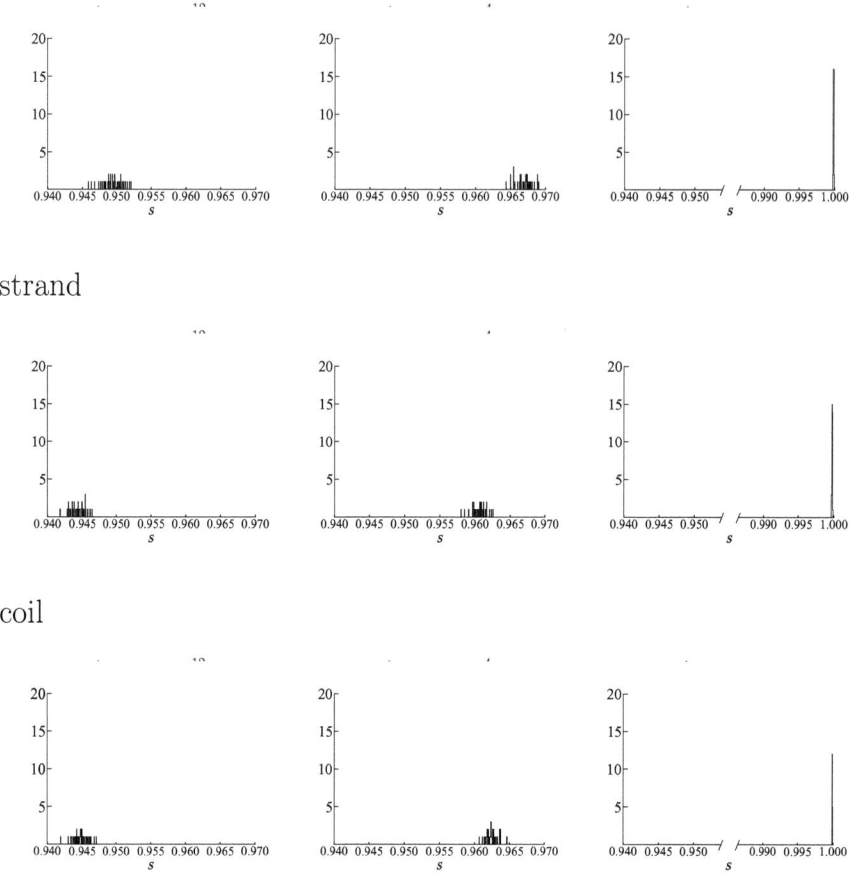

Figure 5.8: Parameter vectors similarity distributions for helix, strand and coil using $\lambda_{\text{quad}} = 10^{-12}$, $\lambda_{\text{quad}} = 10^{-4}$ and $\lambda_{\text{quad}} = 0.9$.

are clearly strongly suppressed and are all forced in a very narrow space as the entity and spread of the similarity factors testify.

Super scoring functions. A similar procedure was subsequently carried out to optimize to the quadratic regularization parameters for the super scoring functions. In this case, larger values of λ_{quad} had only negative influence on the single-choice classification performance. Therefore, nearly vanishing values of $\lambda_{\text{quad}} = 10^{-12}$ were used throughout.

Table 5.1: Multi-choice prediction and recognition performance for helix, strand and coil secondary structure classes (strict DSSP reduction scheme) using respectively (a) a scoring function-based predictor resulting from a pure sequence-structure correlation procedure, (b) a super scoring function-based predictor resulting from a sequence-structure correlation procedure coupled to a structure-structure correlation procedure and (c) a neural network-enhanced predictor resulting from the combination of the sequence-structure and the structure-structure correlation procedures with a third structure-structure correlation procedure carried out by an artificial neural network.

(a) sequence-structure (bare scoring functions)

		prediction		recognition	
	DSSP	$Sens_3$	$Spec_3$	$Sens_3$	$Spec_3$
helix	[H]	0.873 ± 0.005	0.823 ± 0.004	0.879 ± 0.001	0.832 ± 0.001
strand	[E]	0.837 ± 0.004	0.691 ± 0.006	0.849 ± 0.001	0.698 ± 0.001
coil	[C,S,T,G,I,B]	0.734 ± 0.005	0.852 ± 0.005	0.740 ± 0.001	0.859 ± 0.001
$\bar{Q}_3^{(strict)}$		0.801 ± 0.003		0.808 ± 0.001	
$\bar{R}_3^{(strict)}$		0.697 ± 0.005		0.708 ± 0.001	

(b) structure-structure (super scoring functions)

		prediction		recognition	
	DSSP	$Sens_3$	$Spec_3$	$Sens_3$	$Spec_3$
helix	[H]	0.875 ± 0.005	0.845 ± 0.005	0.880 ± 0.001	0.851 ± 0.001
strand	[E]	0.839 ± 0.004	0.710 ± 0.007	0.848 ± 0.001	0.718 ± 0.001
coil	[C,S,T,G,I,B]	0.759 ± 0.006	0.851 ± 0.004	0.765 ± 0.001	0.857 ± 0.001
$\bar{Q}_3^{(strict)}$		0.814 ± 0.004		0.820 ± 0.001	
$\bar{R}_3^{(strict)}$		0.714 ± 0.006		0.724 ± 0.001	

(c) structure-structure (artificial neural network)

		prediction		recognition	
	DSSP	$Sens_3$	$Spec_3$	$Sens_3$	$Spec_3$
helix	[H]	0.868 ± 0.006	0.858 ± 0.008	0.874 ± 0.008	0.864 ± 0.006
strand	[E]	0.760 ± 0.012	0.790 ± 0.012	0.773 ± 0.010	0.800 ± 0.009
coil	[C,S,T,G,I,B]	0.828 ± 0.008	0.821 ± 0.007	0.834 ± 0.009	0.828 ± 0.006
$\bar{Q}_3^{(strict)}$		0.827 ± 0.004		0.834 ± 0.001	
$\bar{R}_3^{(strict)}$		0.727 ± 0.006		0.738 ± 0.001	

5.2.4.2 Estimating performance

With the indications provided by the regularization experiments, the prediction machinery could finally be set up to perform the actual tests on the $\mathcal{R}_{90/10}$ subdivisions. The results achieved by the scoring function-based predictor, respectively at sequence-structure (bare scoring functions) and structure-structure (super scoring functions) correlation level, are reported in tables 5.1a and 5.1b. They show how the longed Q_3-accuracy of 0.8 could finally be overcome, thus proving the usefulness of sequence profiles once again, but also the validity of the method presented here. In the last line of each subtable the average three-class correlation coefficient, R_3 (see figure 2.10), is also reported.

It is of interest to look at the amino acid-specific overall and partial accuracy indices. The amino acid-specific accuracy indices obtained using the super scoring function-based predictor are shown in figures 5.9 and 5.10.

The highest overall Q_3-accuracy is achieved on alanine (A), valine (V), glycine (G) and proline (P): the first two most likely owe their ranking to their predictability in being found respectively in helices and strands (both in the case of valine), as confirmed by helix and strand quality indices in figures 5.10a, 5.10b, 5.10c, 5.10d; the second two probably owe theirs, in contrast, to their tendency to be found outside helices and strands (see coil-specificity and coil-sensitivity in figures 5.10e and 5.10f). The helix and strand classes, are predicted with high specificity and sensitivity also on isoleucine (I), and leucine (L) residues, but the correlation between these amino acids and the coil meta-class seems to be, on the contrary, rather elusive. The helix and strand propensities of methionine (M), glutamine (Q), glutamic acid (E), phenylalanine (F) and tyrosine (Y) is reflected by the specificity and sensitivity values for these residues; asparagine (N) and aspartic acid (D) appear instead to be fairly easy to place within the coil regions. The two amino acids to which it seems most difficult to assign a secondary structure motif are cysteine (C) and tryptophan (W), but this may be in part due also to their low relative abundance (1.5% and 1.4% respectively) in the database.

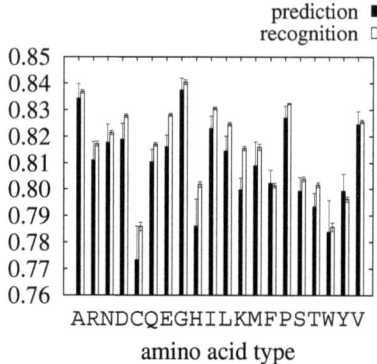

Figure 5.9: Q_3-accuracies (strict DSSP reduction scheme) in prediction and recognition for each amino acid type.

5.2.4.3 Taking a closer look

As anticipated in section 2.1.1.1, a particularly attractive aspect of the model developed is that it enables to easily trace its parameters back to "physical" correlation rules. Let's look then in more detail at the model's internal structure to see if any remarkable features out of all those building up its strength can actually be identified.

Figure 5.11 shows averages and standard deviations of the model's parameters corresponding to the optimal λ_{quad} determined in the tests of section 5.2.4.1.

Some of the known amino acids tendencies mentioned in the introduction are recognizable in the plots 5.11a, 5.11c, and 5.11e relative to the linear parameters. For example, alanine's propensity to be found at the centre of helical patterns is confirmed by the magnitude of the helix scoring function parameters 121, 141

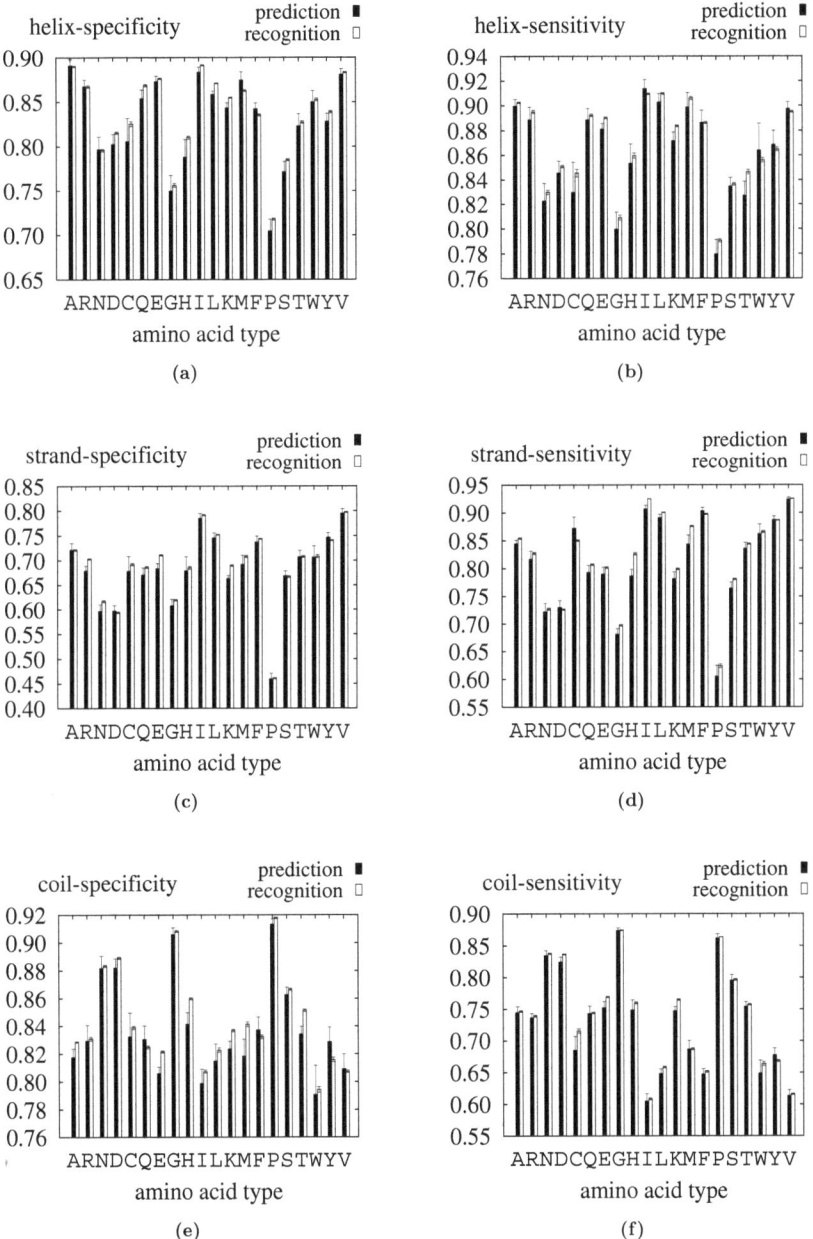

Figure 5.10: Amino acid specific motif-specificities/sensitivities for the three standard secondary structure motifs, helix, strand and coil.

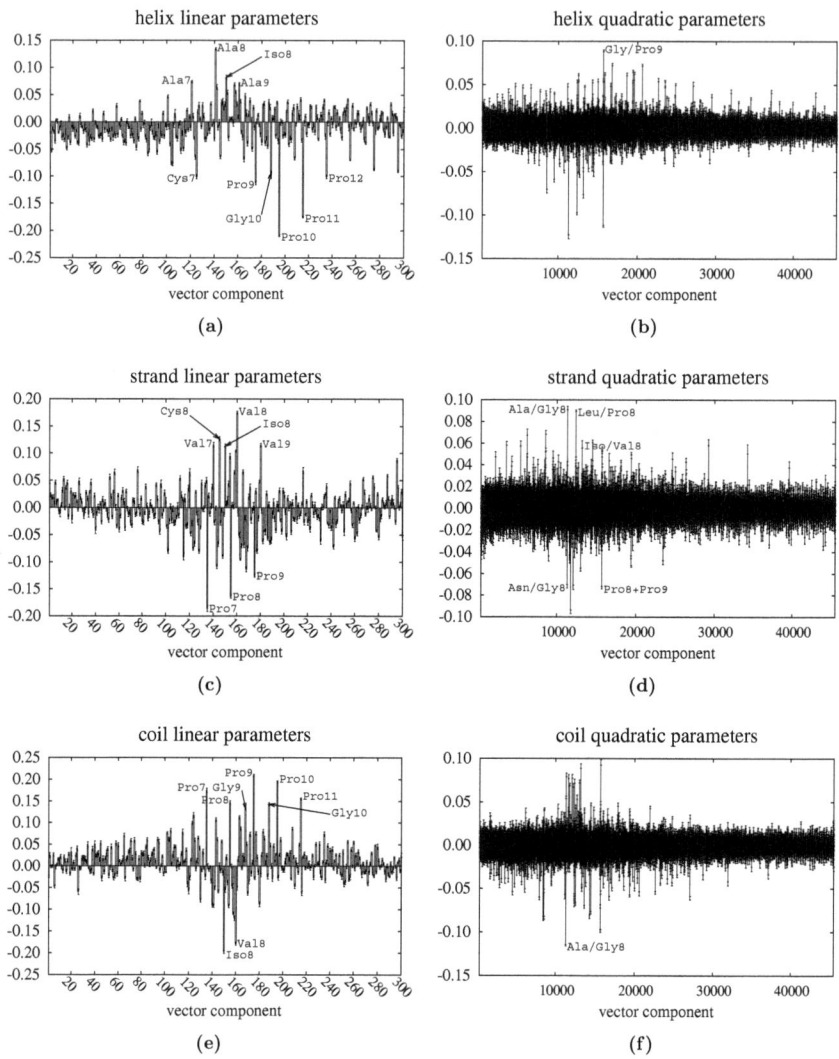

Figure 5.11: Parameter vectors averages and standard deviations for helix, strand and coil obtained using the optimal quadratic regularization parameters. These parameters are the result of the correlation between the sequence window comprising 15 amino acids and the secondary structure pattern of the central (eighth) residue. [Note: the quadratic parameters are sorted as in the vectors (4.2.2), i.e. in the order $W_{1,1}, W_{1,2}, W_{2,2}, W_{1,3}, W_{2,3}, W_{3,3}, \ldots$]

and 161, corresponding to finding it respectively at positions 7, 8 and 9 in the sequence window[4]. Likewise, the presence of valine at positions 7, 8 or 9 in the sequence window, marked by parameters 140, 160 and 180, gives through the strand scoring function, a strong indication that the central residue might be in this conformation. On the other hand, the helix signal is clearly disrupted by the presence of proline at positions 9, 10, 11, as is the strand signal, though more seriously if this residue is found at positions 7, 8 and 9. A relevant structure-breaking effect has of course glycine, but in some positions also cysteine. These amino acids tend to favour instead the coil conformation, as can be seen from the plot 5.11e. Less known tendencies can also be won from the mentioned plots, like the helix/strand ambivalence of isoleucine and a certain propensity of cysteine itself to be in the centre of strands.

But perhaps the most interesting type of information that can be read from figure 5.11 is the one that comes from the plots 5.11b, 5.11d, and 5.11f relative to the quadratic parameters. Although these have in general lower magnitudes than their linear counterparts, it is still possible to discern among them important signal carriers, which amount in this case to second order correlations. Most of the stronger quadratic signals come from *pseudo*-correlations, that is parameters relative to the same residue position in the sequence window. Some of these refer to the same amino acid type as well and therefore play the same role as their linear equivalents. Others, more interesting, refer to different amino acid types and are a consequence of the use of the profile representation. These "quasi-linear" parameters speak for amino acids types inserted in a context which enhances their affinity with other amino acid types. So, for example, a large glycine/proline affinity at position 9 is associated with a favourable signal for helix. While this may be puzzling at first, it makes perfect sense when it is taken into account that the scoring function was optimized to correlate with the central position, that is with position 8. A glycine/proline affinity could then point at the breaking of a helical structure and therefore be a fairly reliable indicator of a helical motif for the preceding residue. Remarkable signals from the strand scoring function include the one coming from double presence of proline at positions 8 and 9, and an isoleucine/valine equivalence at the central position. At the same position are also the more curious alanine/glycine, leucine/proline and asparagine/glycine equivalences to be found: the first two give a positive contribution to the strand likelihood score, the last a negative one. The combinations that do not favour helix or strand conformations favour in general coil.

5.2.4.4 The missing correlations

From the 0.6 Q_3-accuracy (section 4.1) achieved in the early attempts and through the development process described in the present chapter, the scoring

---

[4]Interestingly, there seems to be a marked asymmetry in the magnitude of positive linear parameters indicating a weight shift towards the C-terminus.

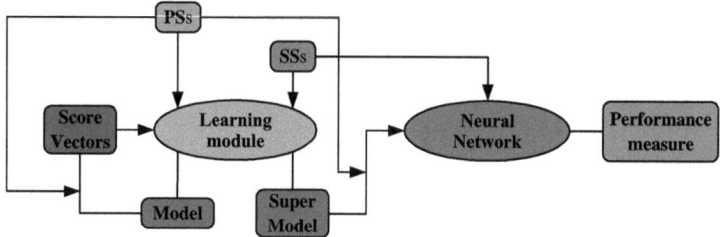

Figure 5.12: Scheme of the neural network-enhanced predictor. The mechanism at the heart the neural network-enhanced predictor is very similar to the one on which the super scoring functions are based. For each residue in the test chains the *super* scores computed for a neighbourhood of that residue are passed to the neural network, which performs its prediction based on them.

function-based secondary structure predictor has come quite a long way.

Introducing second order[5] correlations by including quadratic terms in the bare scoring functions (section 4.2) added a new dimension to the predictor's parametrical power. Using profiles in place of standard sequence vectors on the other hand, gave a decisive boost to the performance by introducing a context-dependent measure of similarity for the amino acid sequences. Finally, The super scoring functions provided correlations between various scores and, in their latest incarnation, also correlations between scores relative to different secondary structure patterns.

In spite of all the upgrades operated however, the predictor's basic structure had not changed: it was still simply a superposition of three independent single-state predictors, with no awareness of each other.

The lacking awareness would be raised in a *third* correlation procedure to be carried out with the help of an artificial neural network [140]. Figure 5.12 shows how the neural network was actually integrated into the predictor's overall architecture.

The neural network employed was a multiple layer perceptron trained by back-propagation. Its input layer consisted of the helix, strand and coil super scores for all the residues in a neighbourhood window of 15 amino acids centred in the target amino acid; the output layer consisted of course in the three final scores for the three secondary structure types of interest; in between them was a hidden layer comprising 100 nodes.

Table 5.1c summarizes the results obtained on the familiar $\mathcal{R}_{90/10}$ subdivisions (excluding the one used to optimize the regularization parameters) using the neural network-enhanced predictor. Its content is to be compared to the one of subtables 5.1a and 5.1b.

---

[5]and fourth order during the optimization procedure.

The most evident effect the neural network had was that of reducing the predictor's sensitivity to helix and especially strand, and to favour in contrast the one to coil. As can be seen by looking also at the specificities reported on table 5.1b, the super scoring functions had a certain tendency to overestimate the strand signal. This tendency was probably detected by the neural network, which took the necessary measures in order to suppress it, benefiting in so doing the sensitivity to coil. Because of the remarkable size of the coil population, the overall Q_3 results profited too by the neural network's intervention.

Part III
applications

Chapter 6
The application software

This chapter is dedicated to the description and the application of the secondary structure prediction software that resulted from the development process described in part II of this report.

6.1 SPARROW

Because of the nature of its basic constituents, that is, the parameter arrays $\tilde{\mathbf{w}}$ (see sections 4.1 and 4.2) and $\tilde{\mathbf{u}}$ (see section 5.2.2), which effectively identify the scoring functions at its core, the prediction software was designated "Secondary structure Predicting ARRays of Optimized Weights", or, in short, SPARROW.

Most of the tests carried out during the development of SPARROW, including the tests of section 5.2.4, made use of protein data extracted from the release 1.71 of the ASTRAL40 data set. While for testing purposes however, part of the data set was always left out for cross-validation, the *whole* data set was used in order to instruct the actual prediction application tool.

The rest of SPARROW's specifications followed exactly those of the program employed in the tests of section 5.2.4.

SPARROW's loose reduction kit. Since in most secondary structure prediction applications the DSSP states are reduced adopting the loose scheme described in section 2.3.1.5, this was adopted in SPARROW's applications as well. In order to compensate for any handicaps that might ensue as a consequence of the fact that it had been optimized using instead the strict reduction scheme, SPARROW was fitted with an *loose reduction kit*. Whenever a residue would be assigned a coil motif, an auxiliary scoring function, optimized to distinguish between exotic (3-turn, 5-turn) helices and other coil types (coil, turn and bend), would be applied. This would ultimately decide whether the residue should indeed be assigned a coil or the prediction should rather be converted to helix[1].

---
[1]Incidentally, the loose reduction kit makes it possible to straightforwardly discriminate between alpha helices and exotic helices.

Table 6.1: SPARROW's overall and partial accuracy in assessing the secondary structure organization of membrane and non-membrane domains, respectively (a) not having and (b) having learned the former.

(a) SPARROW-nm

	DSSP	non-membrane domains (recognition)		membrane domains (prediction)	
		$Sens_3$	$Spec_3$	$Sens_3$	$Spec_3$
helix	[H,G,I]	0.8487	0.8790	0.8070	0.8421
strand	[E,B]	0.7707	0.8107	0.6844	0.7959
coil	[C,S,T]	0.8294	0.7847	0.8089	0.7192
$Q_3^{(loose)}$		**0.8234**		**0.7774**	
$R_3^{(loose)}$		**0.7262**		**0.6590**	

(b) SPARROW-all

	DSSP	non-membrane domains (recognition)		membrane domains (recognition)	
		$Sens_3$	$Spec_3$	$Sens_3$	$Spec_3$
helix	[H,G,I]	0.8621	0.8698	0.8360	0.8486
strand	[E,B]	0.8062	0.7796	0.7766	0.7868
coil	[C,S,T]	0.7976	0.8059	0.7870	0.7704
$Q_3^{(loose)}$		**0.8230**		**0.8018**	
$R_3^{(loose)}$		**0.7268**		**0.6970**	

6.1.1 Introducing transmembrane domains

As the reader may recall from section 4.1.5, the $\mathcal{R}_{90/10}$ data subdivisions employed in the software development were generated after exclusion of the transmembrane domains present in the original ASTRAL40 data set. When instructing the first version of SPARROW this tradition was maintained and the membrane domains were left out of the so-called knowledge-data set provided to the learning module. The question later arose, whether this was indeed a wise choice or not. A second version was therefore instructed with all domains in the original data set, *including* the transmembrane domains. Its performance was compared to that of the original one in the recognition of the learned data set and in the prediction/recognition of the portion comprising only transmembrane domains.

The results in table 6.1 indicate on the one hand that the secondary structure of transmembrane domains is indeed somewhat harder to predict than that of other domains. On the other hand, they quite undoubtedly show that, while it of course benefits the performance achievable on the transmembrane domains themselves, the inclusion of these in the learning process has practically no repercussion on the prediction of other domains.

In light of the indications provided by these tests, the version instructed with the complete ASTRAL40 data set was used in all applications thereafter.

Table 6.2: SPARROW-all's overall and partial accuracy on photosystems I and II.

	DSSP	PS I		PS II	
		$Sens_3$	$Spec_3$	$Sens_3$	$Spec_3$
helix	[H,G,I]	0.7952	0.8879	0.6966	0.9154
strand	[E,B]	0.5533	0.3897	0.5900	0.2986
coil	[C,S,T]	0.7801	0.7229	0.7694	0.7053
$Q_3^{(loose)}$		0.7733		0.7182	
$R_3^{(loose)}$		0.6006		0.5457	

6.1.1.1 Two special membrane proteins

Because of their particular relevance for the Earth's biosphere, the two major protein complexes responsible for photosynthesis, the photosystems I and II, were also looked into, to see how well the fully trained SPARROW could predict their secondary structure. The results are collected in table 6.2.

As can be seen, the secondary structure is not predicted with very high accuracy in either photosystem and especially in photosystem II. Besides being membrane protein, which already makes them potentially tough candidates, the photosystems contain also a great number of cofactors (chlorophylls, carotenoids), which are taken in no account by SPARROW but could play a role in the structure formation as well.

6.1.2 Prediction confidence

The Q_3-accuracy, estimated by means of cross-validation techniques like the one employed in the final benchmark of section 5.2.4, does provide a general measure of a predictor's reliability. However, different secondary structure types are predicted with different accuracies for different amino acids, and sometimes it could be desirable to have a more detailed picture of the reliability of a prediction. One might for example be interested in the secondary structure prediction for the amino acids well within a peptide chain and not for those near the terminal positions. For this reason it is important to provide, along with a residue-wise secondary structure prediction, also a measure of its *confidence*.

The natural source of information out of which to build the prediction confidence are the super scores F_σ (or alternatively, the neural network output signals for the secondary structures of interest) for every residue in the target protein chain. At first sight, it would seem quite reasonable to assume that the larger one score is as compared to the others, the more reliable the ensuing prediction becomes. It is however imaginable also that certain situations might arise in which the predictor is badly mistaken. On all accounts, its ways could very well be too difficult to fathom for heuristic human-devised "magic formulas".

Since the machinery developed in the course of the project was actually de-

Table 6.3: Expected prediction accuracies (in percent) for all confidence levels. The large fluctuations reported for confidence levels 1 and 9 are a consequence of the small number of residues the secondary structure of which was predicted with such a confidence (see figure 6.2a).

confidence level	helix(%)	strand(%)	coil(%)	overall(%)
1	(49.7 ± 52.3)	(40.0 ± 46.5)	(60.8 ± 27.8)	(53.1 ± 25.5)
2	(54.2 ± 5.9)	(45.4 ± 4.4)	(66.3 ± 1.9)	(57.7 ± 1.5)
3	(66.0 ± 2.5)	(56.6 ± 1.9)	(74.8 ± 2.2)	(67.5 ± 1.3)
4	(77.0 ± 1.9)	(72.4 ± 3.2)	(81.6 ± 0.9)	(77.9 ± 1.1)
5	(86.6 ± 2.0)	(84.5 ± 2.6)	(86.5 ± 1.5)	(86.0 ± 1.4)
6	(92.7 ± 1.9)	(91.9 ± 1.5)	(90.5 ± 2.4)	(91.8 ± 1.1)
7	(95.9 ± 1.3)	(95.2 ± 1.8)	(93.4 ± 1.1)	(95.1 ± 0.6)
8	(97.9 ± 1.8)	(97.4 ± 2.9)	(95.8 ± 3.1)	(97.3 ± 2.1)
9	(98.9 ± 8.9)	(97.9 ± 29.6)	(97.8 ± 10.5)	(98.6 ± 6.8)

signed to distinguish between two classes ("right" and "wrong" perhaps?), the confidence determination problem looked very much like an application which a scoring function is perfectly well suited for.

The confidence scoring function C would be a function of the final secondary structure scores for the target residue. Given the secondary structure scores s_σ, $\sigma \in \mathcal{P}$, it will be

$$C(\mathbf{g}, g_0; \mathbf{s}) = \mathbf{g} \cdot \mathbf{s} + g_0, \quad \mathbf{g} \in \mathbb{R}^m, \ g_0 \in \mathbb{R}, \qquad (6.1.1)$$

where $m = |\mathcal{M}|$ is the number of secondary structure motifs of interest, and all scores s_σ are grouped in the vector $\mathbf{s}$.

Like all scoring function analysed so far, also the confidence function $C(\mathbf{s})$ would be optimized to be close to one or zero, one representing this time the value labelling the "right", zero the value labelling the "wrong" predictions.

SPARROW's confidence scoring function was optimized on the whole knowledge data set and was applied in turn to every $\mathcal{R}_{90/10}$ subset. As experienced with other scoring functions optimized in this way, also in this case the returned confidence values were only roughly in the range [0, 1]. For simplicity, the whole confidence range was divided into nine intervals and each interval was assigned a digit between 1 and 9. During the test prediction the residues in the chains were sorted according to the confidence their secondary structure was predicted with, and the actual specificities were recorded for each confidence interval. Figure 6.1 reports the average confidence-dependent populations and specificities for the three standard secondary structure types, while figure 6.2 shows the corresponding global quantities.

The specificities ultimately embody the probability that a prediction of the pertaining secondary structure be correct, and can be a useful aid in the prediction software applications (see section 6.2.4). The expected prediction accuracies are reported for easier reference also in table 6.3.

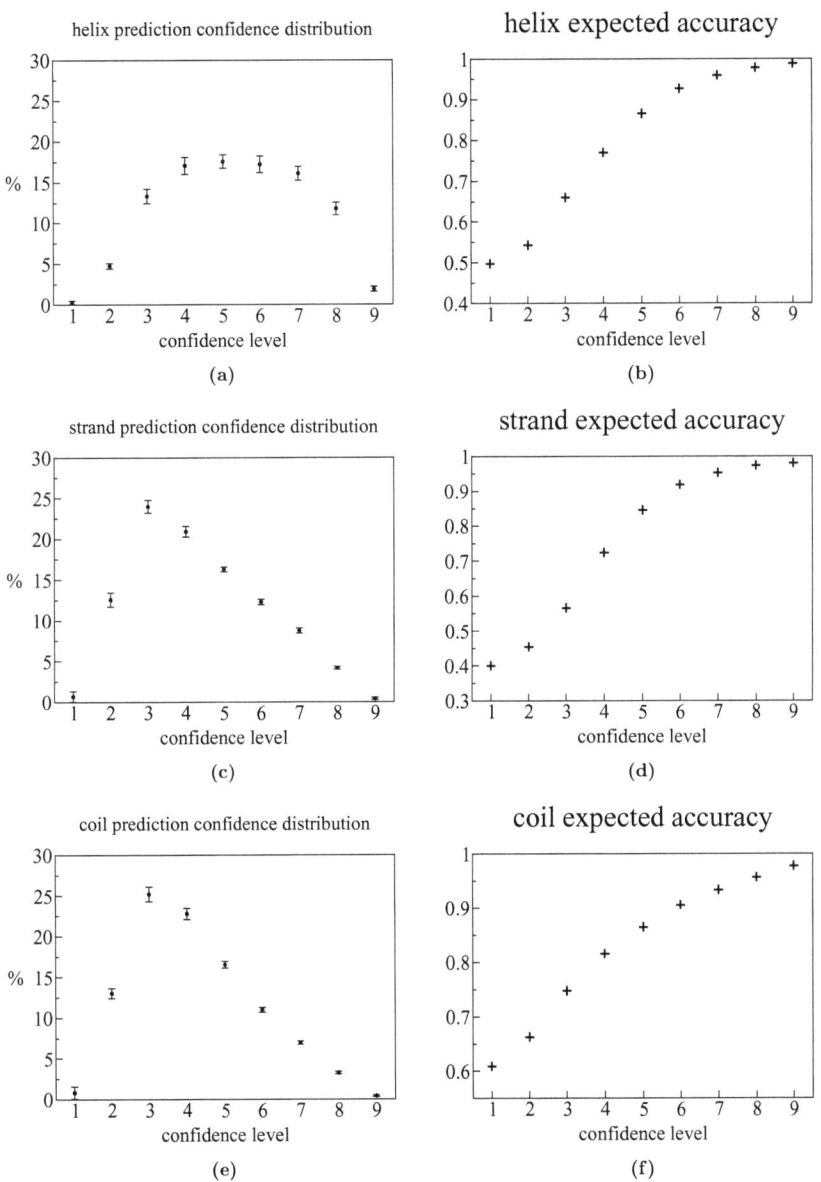

Figure 6.1: The plots in (a), (c) and (e) show the average percentage of residues whose secondary structure was predicted with the corresponding confidence level; those in (b), (d) and (f) show the expected accuracy for the prediction of the relative secondary structure, that is, the probability that that prediction be actually correct.

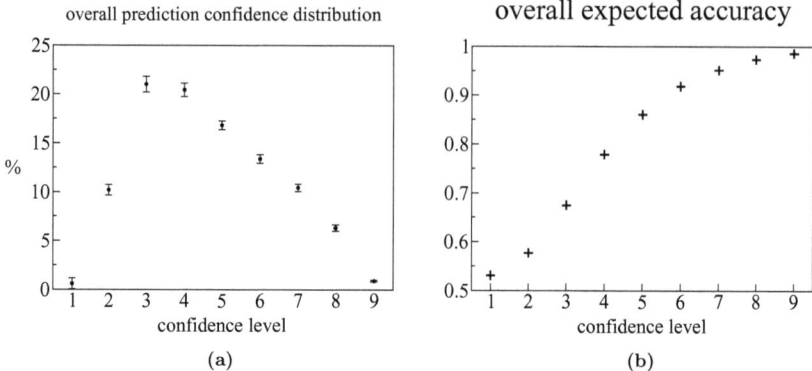

Figure 6.2: These plots are similar to the ones in figure 6.1, but do not refer to one secondary structure in particular. They can be of use if the *general* accuracy of the secondary structure prediction is needed.

6.2 SPARROW versus others

In order to pronounce the final verdict on the quality of SPARROW's secondary structure prediction, this must be compared to the one achievable using other secondary structure prediction programs. The designated terms of comparison were selected among the programs available for download at the time of writing, and included PROF v1.0 [90], Prospect 2 [101], PSIPRED v2.5 [51], SSpro 4.01 [87], GOR-IV [42], PHD [100] and PREDATOR [73].

6.2.1 ASTRAL40 generations: a neutral testing ground

A neutral testing ground had to be set for the comparison to be a fair one.

While SPARROW was under development the version 1.73 of the ASTRAL compendium was released. Stemming from the SCOP database, this data set was very likely to contain most of the structures already present in the one used to instruct SPARROW, or at least very similar ones. A thorough filtering procedure was therefore needed in order to prevent the introduction of biases in the prediction accuracy results. For this purpose the BLAST [141] alignment tool was employed to match every sequence in ASTRAL40 – release 1.73 to the sequences in ASTRAL40 – release 1.71. All the ones which were assigned more than 40% sequence identity (see figure 6.3) were discarded to assemble a reasonably neutral test data set. The BLAST filtering returned 1919 domains.

The overall Q_3-accuracies (loose DSSP reduction scheme) of SPARROW and its competitors on the new test data set are reported in the first column of table 6.4 (the complete results for SPARROW are reported in appendix B). Since their performance lay well below that of other predictors, PHD, PREDATOR

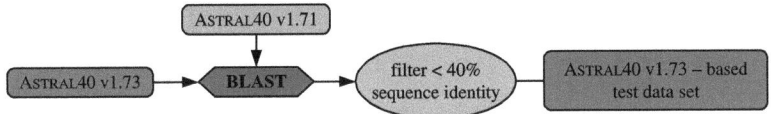

Figure 6.3: Scheme of the BLAST-operated filtering of the ASTRAL40 – release 1.73 data set. The domains in the release 1.73 were pairwise aligned to those in the release 1.71 using the sequence alignment tool BLAST. The domains that, according to BLAST, presented more than 40% sequence identity to those in the release 1.71 were discarded. The remaining ones were used as benchmark data set.

and GOR-IV were excluded from the final report.

As the first objective measure of its efficacy in the secondary structure prediction, the outcome of this test confirms SPARROW as a valid alternative to the existing software. A detailed analysis of the prediction outcome reveals furthermore that the sets of secondary structure motifs wrongly predicted by PSIPRED and SPARROW overlap by only 70.3%. This leaves a non-negligible margin for improvement in a *combined* method.

It is instructive to look at how the performances of the three programs compared in relation to the degree of similarity (always according to BLAST) of the protein domains to the ones learned by SPARROW. Four possible similarity-sorting criteria were identified:

- percentage of sequence identity;

- percentage of residues positively contributing to the BLAST score;

- BLAST score;

- p-value.

The p-value is a measure of the probability that a better alignment, that is, an alignment with a higher BLAST-score, could be found by means of a random database search.

The sorted results are reported in figures 6.4 and 6.5 (the corresponding generalized correlation coefficient and partial helix, strand and coil accuracy indices for SPARROW are again reported in appendix B). Interestingly, they seem to reveal nearly no correlation between the performance quality and the percentage of identity (fig. 6.4a) or similarity (fig. 6.4b) among learned and predicted sequences. On the contrary, in some cases the performance quality seems to be even anticorrelated to the sequence identity (fig. 6.5a). Much more reliable similarity measures are instead the BLAST score and the p-value. When the domains are sorted according to these (see figures 6.4c and 6.4d) it becomes in fact possible to discern a slight decrease in the prediction accuracy. This is

Table 6.4: Q_3-accuracy (loose DSSP reduction scheme) of SPARROW, PROF [90], Prospect [101], PSIPRED [51] and SSpro4 [87] in predicting the secondary structure of the protein domains in ASTRAL40 – release 1.73 with up to 40% sequence identity to those in ASTRAL40 – release 1.71. [Note: SPARROW's reference Q_3-accuracy in the recognition of ASTRAL40 – release 1.71 amounts to 0.8224.]

PROGRAM	$Q_3^{(loose)}$ [ASTRAL40 – release 1.73]	
	all sequences	non-aligned sequences(*)
PROF	0.7508	0.6769
Prospect	0.7748	0.6982
PSIPRED	0.8194	0.7317
SPARROW	**0.8089**	**0.7498**
SSpro4	0.7923	0.7473

(*) these are sequences that returned no significant alignment to the domains in ASTRAL40 – release 1.71

very much enhanced in the cumulative Q_3-accuracy figures 6.5c and 6.5d. Even these plots show a more or less extended transition in the range corresponding to the sequences for which a relatively close match was found in the learned data set. This however, could well be an artifact of the lower statistical weight of the latter (see population plots).

It is worth underlining here how all tested programs (with the possible exception of SSpro4) present a very similar behaviour, in spite of the fact that the data set which the sequences in ASTRAL40 – release 1.73 were matched against, were those learned by SPARROW. This means that the knowledge data of the other predictors could not be too dissimilar from the release 1.71 of ASTRAL40.

The sorted plots confirm the ranking established for the overall performances in table 6.4. This may not be very evident from the partial results in figure 6.4, but it becomes by looking at the cumulative results collected in figure 6.5.

6.2.2 Non-aligned sequences

The sequences sorted in the plots in figures 6.4 and 6.5 are not all those resulting from the BLAST filtering procedure: 22 out of the 1919 (for a total of 1379 residues) did not return any reasonable BLAST alignments and could not be included in such sorting[2].

The performance of the tested programs on these outliers is reported in the second column of table 6.4 (partial results for SPARROW are as usual reported in appendix B). These results are more or less in line with those achieved on the "aligned" domains. They however see SPARROW this time in the lead by a short measure over SSpro4. Even if the size of the data set accords them only very limited statistical weight, these results speak in favour of SPARROW's generalization capability.

---

[2]A closer inspection of the profiles relative to these sequences revealed no anomalies.

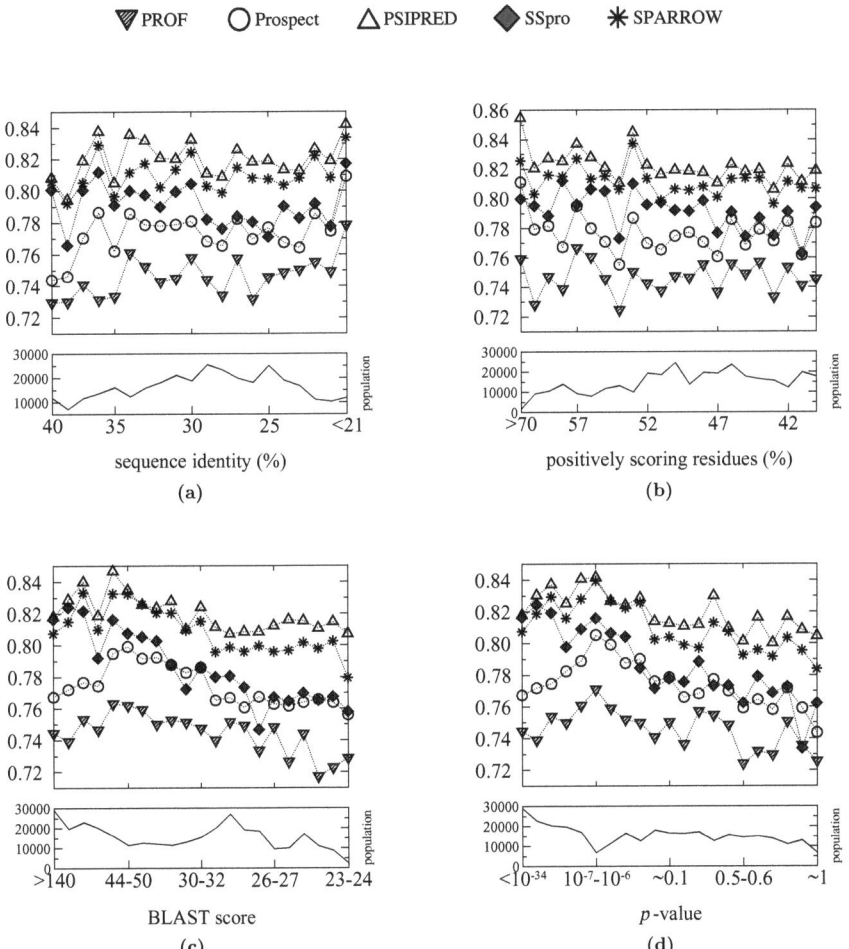

Figure 6.4: Q_3-accuracy (loose DSSP reduction scheme) achieved by SPARROW, PROF, Prospect, PSIPRED and SSpro4 on the 1919 domains of ASTRAL40 – release 1.73 with less than 40% sequence homology to the domains of ASTRAL40 – release 1.71. The domains are sorted here respectively according to (a) percentage of sequence identity, (b) percentage of residues that positively contributed to the BLAST alignment score, (c) BLAST alignment score and (d) p-value. The partial populations (number of residues) in the ASTRAL40 – release 1.73 data set are reported in the lower portion of the plots.

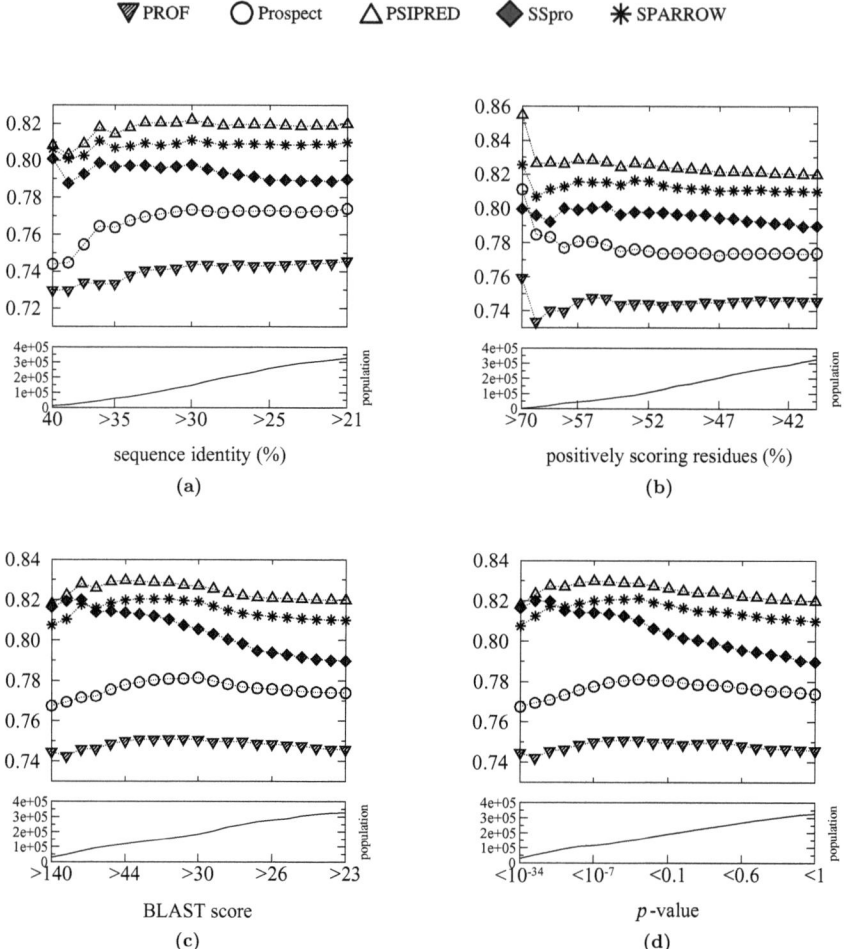

Figure 6.5: Q_3-accuracy (loose DSSP reduction scheme) achieved by SPARROW, PROF, Prospect, PSIPRED and SSpro4 on increasingly greater portions of the 1919 domains of ASTRAL40 – release 1.73 with less than 40% sequence homology to the domains of ASTRAL40 – release 1.71. These portions were obtained by progressively adding chains with respectively (a) lower percentage of sequence identity, (b) lower percentage of residues that positively contributed to the BLAST alignment score, (c) lower BLAST alignment score and (d) higher p-value. The cumulative populations, in terms of number of residues, are reported in the lower portion of the plots.

6.2.3 A 3D picture of secondary structure prediction

In order to give a different flavour of SPARROW's secondary structure prediction ability, 23 domains of ASTRAL40 – release 1.73 were selected out of the 1919 tested for a three-dimensional visualization. They are representatives of different families and include eight mainly alpha domains, eight mainly beta domains and seven mixed alpha/beta domains. The drawings are collected in the following pages (figures 6.6, 6.7, 6.8, 6.9, 6.10 and 6.11). The correctly predicted secondary structure motifs are coloured, while the non-correctly predicted ones are transparent. The molecules were drawn with the help of the visualization tool VMD.

(a) **1t72** (A) $\{Q_3^{(\text{loose})} = 0.9163\}$ (b) **1wdy** (A) $\{Q_3^{(\text{loose})} = 0.8140\}$

(c) **1x9f** (B) $\{Q_3^{(\text{loose})} = 0.9310\}$ (d) **1xqo** (A) $\{Q_3^{(\text{loose})} = 0.7866\}$

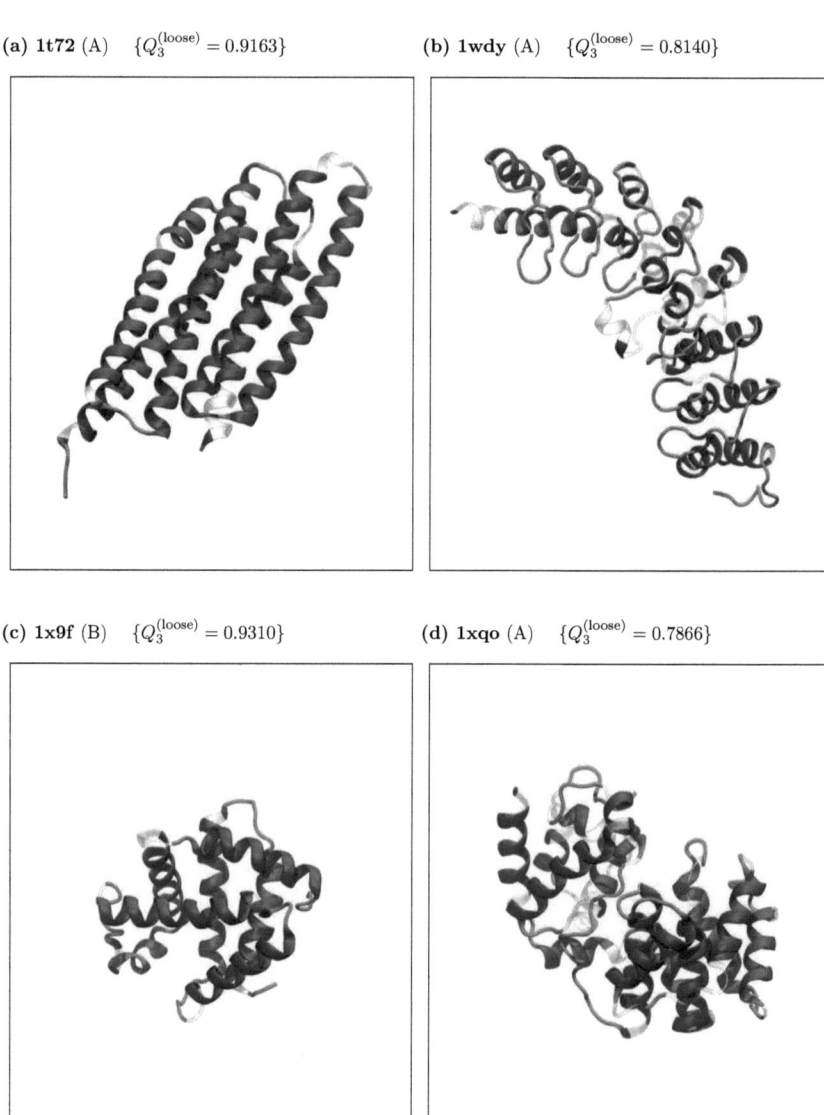

Figure 6.6: Three-dimensional representation of domains with prevalent helix content. In (a) phosphate transport system protein (PhoU) from Aquifex aeolicus, (b) human 2-5a-dependent ribonuclease, (c) subunit II of extracellular dodecameric hemoglobin (erythrocruorin) from Lumbricus terrestris, (d) 8-oxoguanine DNA glycosylase (AgoG) from Pyrobaculum aerophilum.

(a) **1xqr** (A1) $\{Q_3^{(\text{loose})} = 0.8561\}$

(b) **1z0j** (B1) $\{Q_3^{(\text{loose})} = 0.8431\}$

(c) **2aho** (B1) $\{Q_3^{(\text{loose})} = 0.7912\}$

(d) **2c7n** (A1) $\{Q_3^{(\text{loose})} = 0.8182\}$

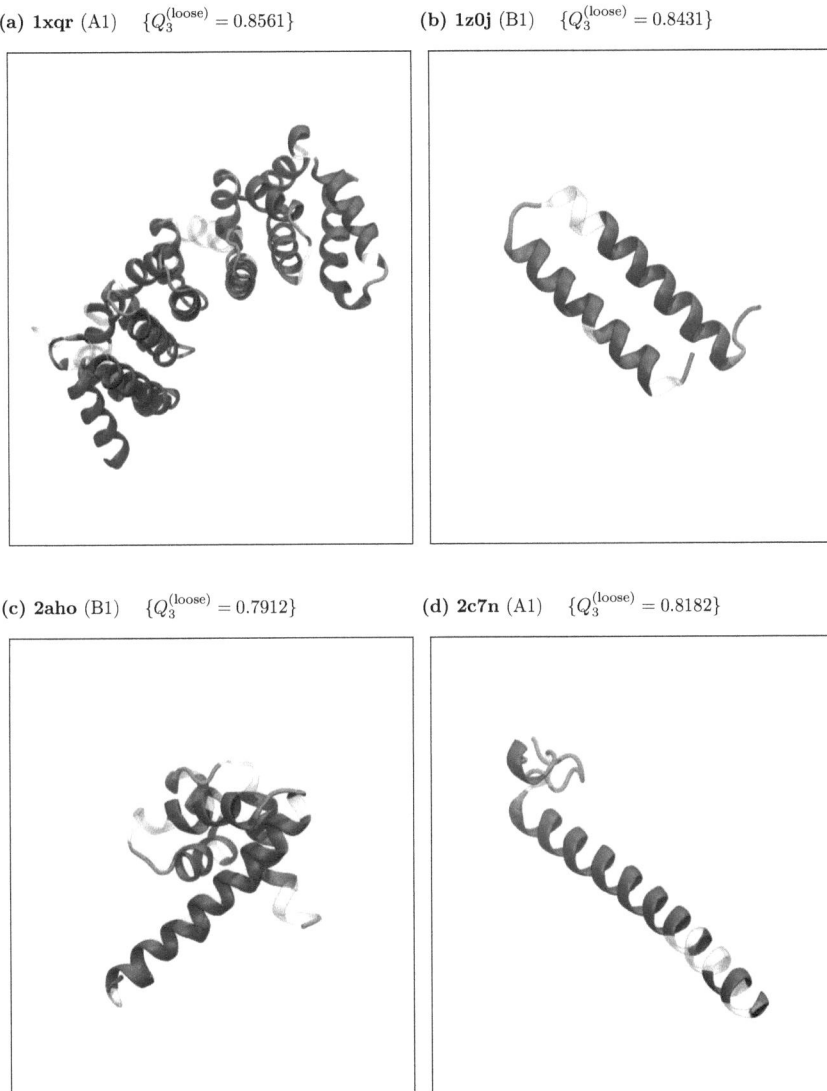

Figure 6.7: Three-dimensional representation of domains with prevalent helix content. In (a) human Hsp70-binding protein 1 (HspBP1), (b) human effector protein rabenosyn-5 (Rab5), (c) domain 2 of the eukaryotic initiation factor 2α (eIF2α) from Sulfolobus solfataricus, (d) ubiquitin-binding domain of human RabGEF1 (Rabex-5).

(a) **1t9h** (A1) $\{Q_3^{(\text{loose})} = 0.8358\}$ (b) **1w9s** (A) $\{Q_3^{(\text{loose})} = 0.8358\}$

(c) **1wqw** (A1) $\{Q_3^{(\text{loose})} = 0.5957\}$ (d) **1y7b** (A2) $\{Q_3^{(\text{loose})} = 0.8349\}$

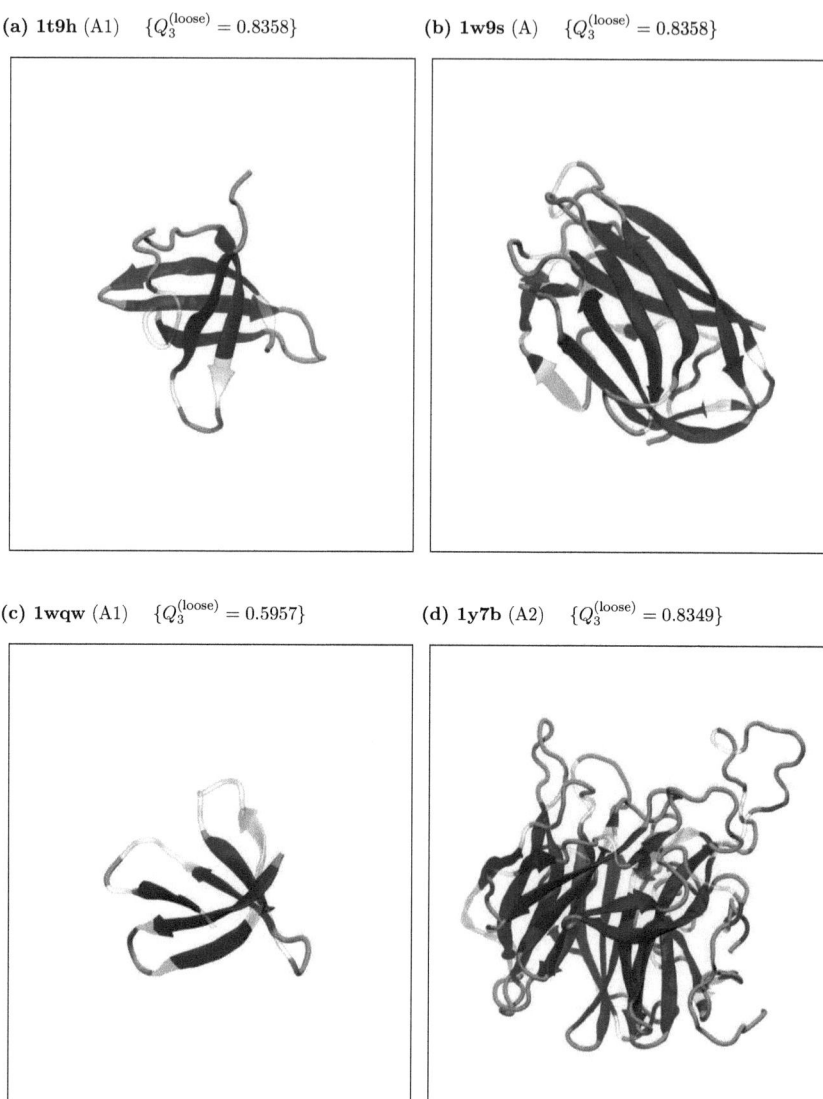

Figure 6.8: Three-dimensional representation of domains with prevalent strand content. In (a) N-terminal domain of probable GTPase EngC (YjeQ) from Bacillus subtilis, (b) hypothetical protein BH0236 from Bacillus halodurans, (c) C-terminal domain of biotin–[acetyl-CoA-carboxylase] ligase from Archaeon Pyrococcus horikoshii, (d) N-terminal domain of beta-D-xylosidase from Clostridium acetobutylicum.

(a) **1y9q** (A2) $\{Q_3^{(\text{loose})} = 0.8485\}$ (b) **2bib** (A1) $\{Q_3^{(\text{loose})} = 0.8017\}$

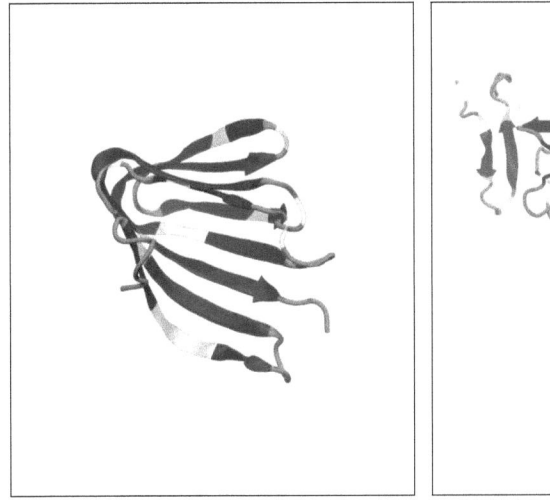
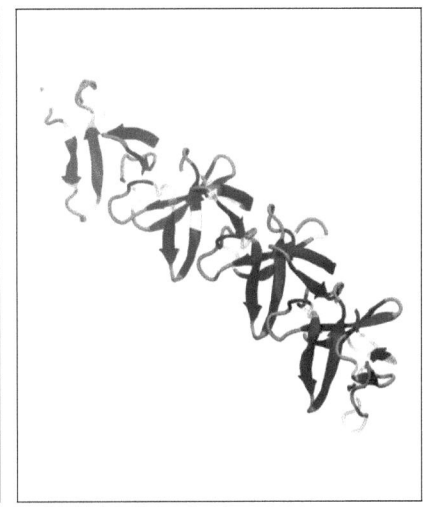

(c) **2f9c** (A1) $\{Q_3^{(\text{loose})} = 0.6406\}$ (d) **2gtl** (O1) $\{Q_3^{(\text{loose})} = 0.8525\}$

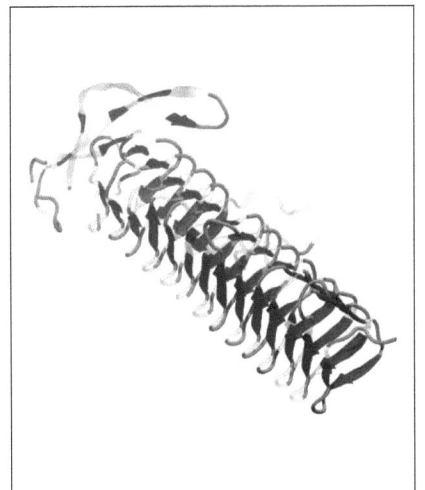
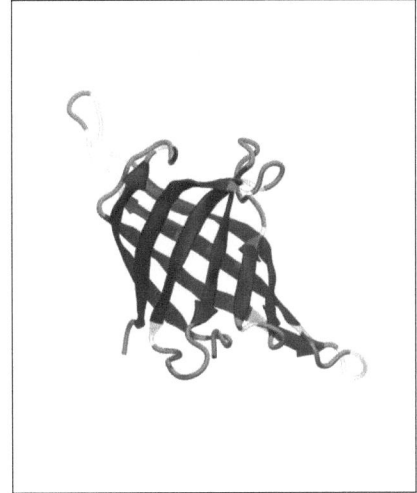

Figure 6.9: Three-dimensional representation of domains with prevalent strand content. In (a) C-terminal domain of probable transcriptional regulator VC1968, from Vibrio cholerae, (b) C-terminal domain of teichoic acid phosphorylcholine esterase Pce (LytD) from Streptococcus pneumoniae, (c) hypothetical protein YdcK from Salmonella enterica, (d) extracellular hemoglobin linker l3 subunit from Lumbricus terrestris.

(a) 1u5h (A) $\{Q_3^{(\text{loose})} = 0.9103\}$ **(b) 1yks** (A1) $\{Q_3^{(\text{loose})} = 0.8429\}$

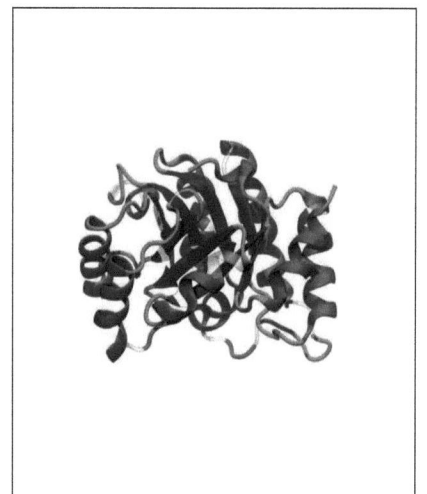
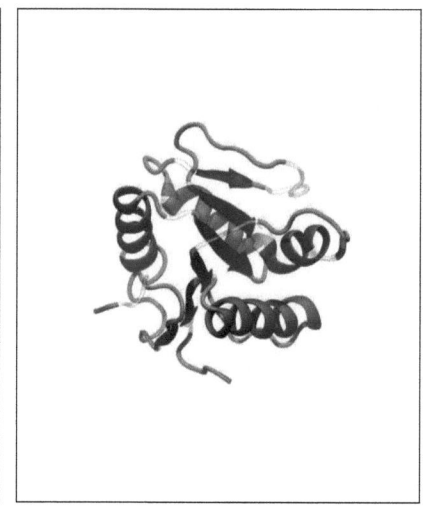

(c) 1yln (A1) $\{Q_3^{(\text{loose})} = 0.7658\}$ **(d) 1zjc** (A1) $\{Q_3^{(\text{loose})} = 0.8039\}$

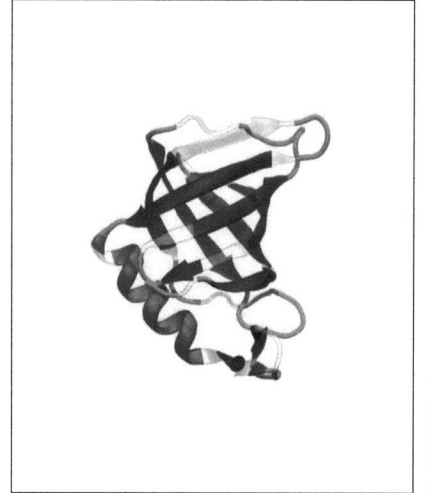
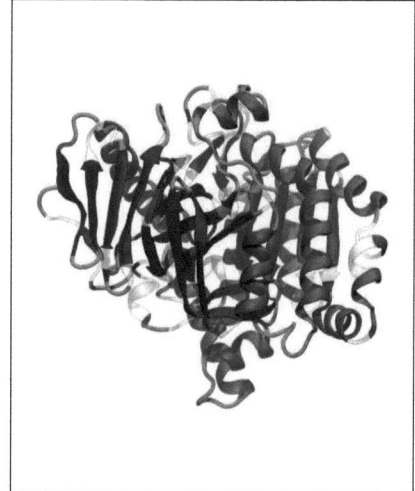

Figure 6.10: Three-dimensional representation of tested domains with mixed secondary structure content. In (a) beta subunit of citrate lyase from Mycobacterium tuberculosis, (b) NS3 helicase from yellow fever virus, (c) protein VCA0042N from Vibrio cholerae, (d) aminopeptidase S (AMPS) from Staphylococcus aureus.

(a) **2g5f** (A1) $\{Q_3^{(\text{loose})} = 0.7770\}$

(b) **2hu7** (A1) $\{Q_3^{(\text{loose})} = 0.7987\}$

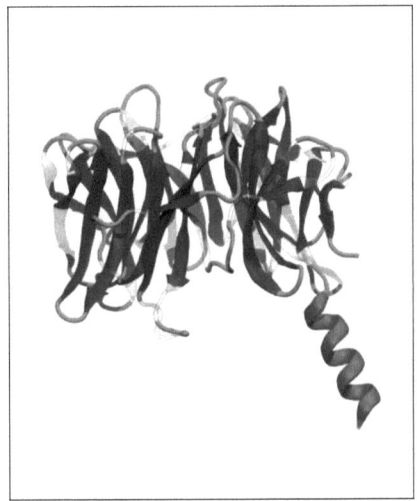

(c) **2j8k** (A1) $\{Q_3^{(\text{loose})} = 0.8114\}$

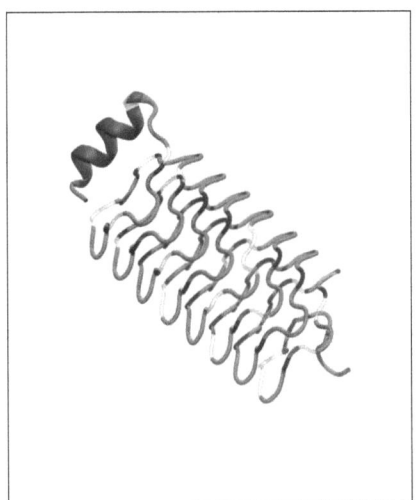

Figure 6.11: Three-dimensional representation of two domains with mixed secondary structure content and one with little secondary structure content. In (a) salicylate synthase MbtI from Mycobacterium tuberculosis, (b) acylamino-acid-releasing enzyme from Aeropyrum pernix, (c) NP275-NP276 from Nostoc punctiforme.

6.2.4 SPARROW's output examples

As mentioned in section 2.1.1.2, the secondary structure predictor assigns a secondary structure motif to every residue in a peptide chain. Sometimes though, one may wish to have a somewhat deeper insight into the secondary structure assignment.

A feature offered by several modern secondary structure predictors is that of integrating the actual output prediction string with a plot depicting the trend of the secondary structure signals along the whole sequence.

This feature is implemented in SPARROW as well. It is illustrated here for eight of the 1919 ASTRAL40 – release 1.73 domains tested. The likelihood scores produced by SPARROW's super scoring functions were plotted as a function of the residue position. Since it was in this case available, the "true secondary structure" (according to DSSP) is reported in the abscissa. The accuracy indices are also reported for possible reference. The plots are collected in pages 122 to 129. They are followed by the actual secondary structure prediction and its confidence level (indicated by colour-code). The reader may refer to table 6.3 and figures 6.1 and 6.2 for the expected prediction accuracy corresponding to each confidence level.

1t72 Phosphate transport system protein (PhoU) from Aquifex aeolicus (see figure 6.6a). This mainly helical structure is predicted with high accuracy. This domain scored 124.0 (with a p-value of $4 \cdot 10^{-30}$) against the domains in the release 1.71 of ASTRAL40.

1wdy Human 2-5a-dependent ribonuclease (see figure 6.6b). While the overall prediction is quite accurate that of the central region, in particular between residues 118 and 173, has a very low confidence value. A brief glance at the corresponding region in the score plot shows how this clearly is a consequence of the closeness of the three super scores for helix, strand and coil. This domain scored 90.5 (with a p-value of $8 \cdot 10^{-20}$) against the domains in the release 1.71 of ASTRAL40.

2f9c Hypothetical protein YdcK from Salmonella enterica (see figure 6.9c). This structure consists of a series of short strands forming a rod of triangular section (see also figure 6.9c). SPARROW does identify all strands but tends in most cases to overestimate their length. This domain scored 32.7 (with a p-value of 0.02) against the domains in the release 1.71 of ASTRAL40.

2gtl Extracellular hemoglobin linker l3 subunit from Lumbricus terrestris (see figure 6.9d). This domain scored 25.4 (with a p-value of 0.02) against the domains in the release 1.71 of ASTRAL40.

1yks NS3 helicase from yellow fever virus (see figure 6.10b). This domain scored 23.5 (with a p-value of 0.99) against the domains in the release 1.71 of ASTRAL40.

1yln VCA0042 from Vibrio cholerae (see figure 6.10c). This domain scored 23.5 (with a p-value of 0.94) against the domains in the release 1.71 of ASTRAL40.

2hu7 N-terminal domain of acylamino-acid-releasing enzyme from Aeropyrum pernix (see figure 6.11b). This domain scored 28.1 (with a p-value of 0.44) against the domains in the release 1.71 of ASTRAL40.

2j8k NP275-NP276 from Nostoc punctiforme (see figure 6.11c). The general low confidence contrasts with the relatively good Q_3-accuracy achieved. This domain scored 27.7 (with a p-value of 0.29) against the domains in the release 1.71 of ASTRAL40.

Note: The following pages are best viewed in colour. For the interested reader a colour version of the original thesis is provided by the FU Berlin online library web service at http://www.ub.fu-berlin.de.

1t72 (A)

helix [H,G,I]		strand [E,B]		coil [C,S,T]		$Q_3^{(loose)}$	$R_3^{(loose)}$
Sens₃	Spec₃	Sens₃	Spec₃	Sens₃	Spec₃		
0.927	0.978	–	0.000	0.833	0.690	**0.916**	**0.659**

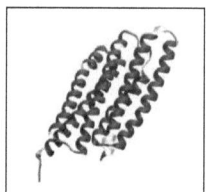

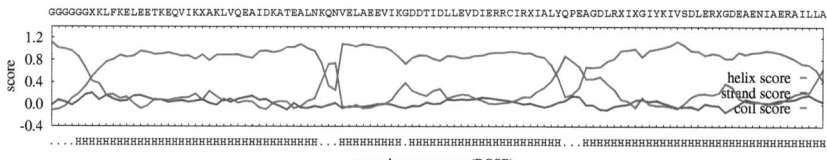

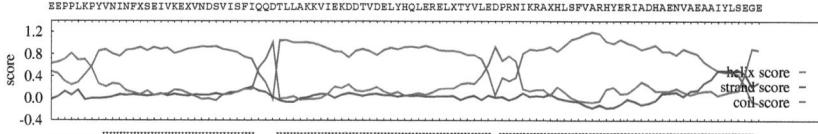

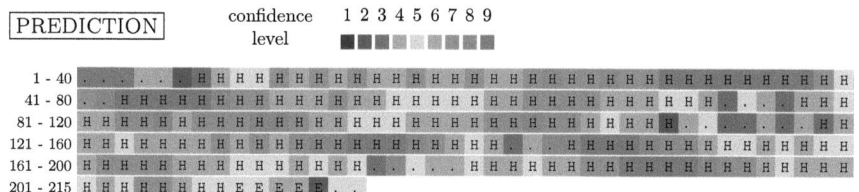

1wdy (A)

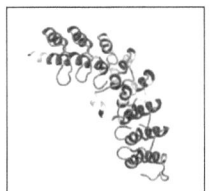

helix [H,G,I]		strand [E,B]		coil [C,S,T]		$Q_3^{(loose)}$	$R_3^{(loose)}$
Sens$_3$	Spec$_3$	Sens$_3$	Spec$_3$	Sens$_3$	Spec$_3$		
0.762	0.893	–	0.000	0.879	0.778	**0.814**	**0.643**

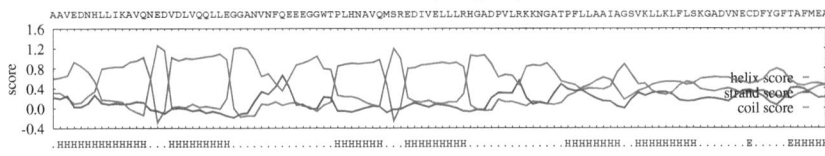

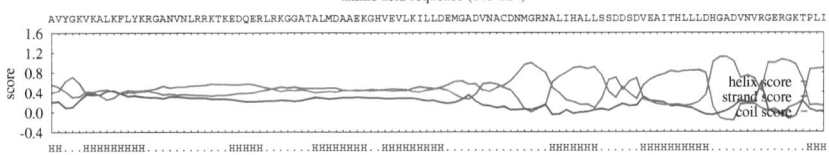

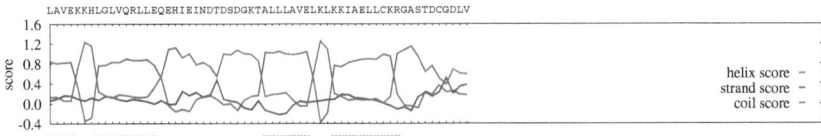

2f9c (A1)

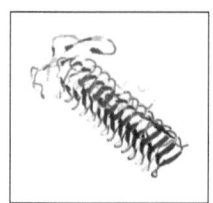

helix [H,G,I]		strand [E,B]		coil [C,S,T]		$Q_3^{(loose)}$	$R_3^{(loose)}$
Sens$_3$	Spec$_3$	Sens$_3$	Spec$_3$	Sens$_3$	Spec$_3$		
–	0.000	0.822	0.549	0.518	0.868	**0.641**	**0.374**

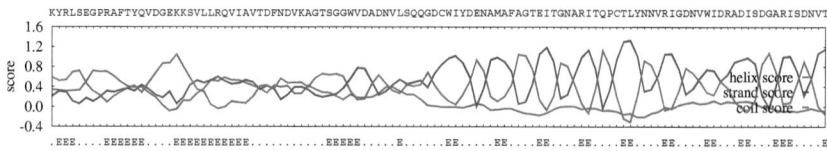

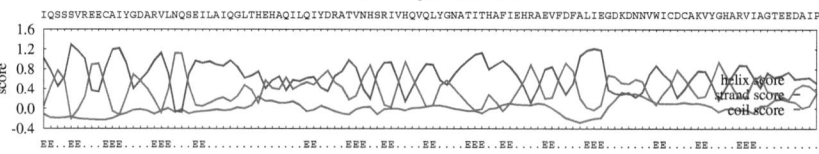

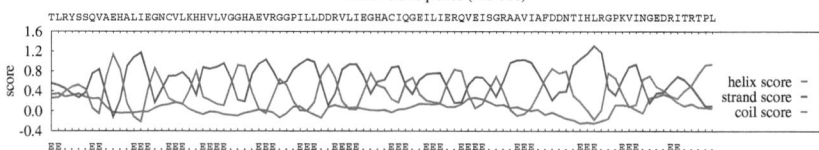

2gtl (O1)

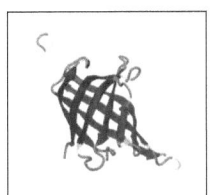

helix [H,G,I]		strand [E,B]		coil [C,S,T]		$Q_3^{(loose)}$	$R_3^{(loose)}$
Sens$_3$	Spec$_3$	Sens$_3$	Spec$_3$	Sens$_3$	Spec$_3$		
—	0.000	0.930	0.880	0.745	0.884	**0.852**	**0.706**

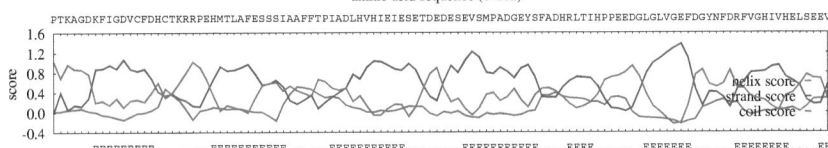

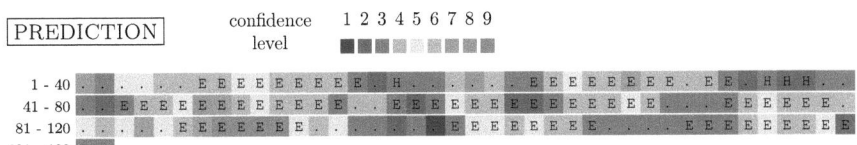

1yks (A1)

helix [H,G,I]		strand [E,B]		coil [C,S,T]		$Q_3^{(loose)}$	$R_3^{(loose)}$
Sens$_3$	Spec$_3$	Sens$_3$	Spec$_3$	Sens$_3$	Spec$_3$		
1.000	0.917	0.889	0.585	0.725	0.980	**0.843**	**0.777**

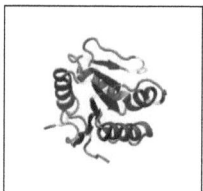

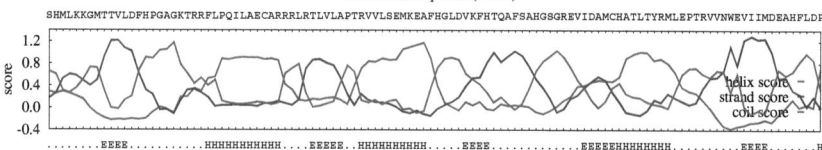

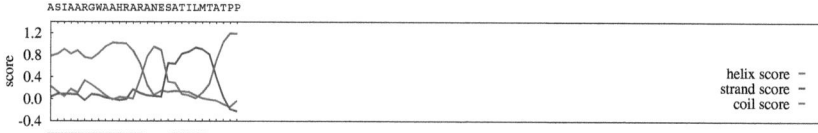

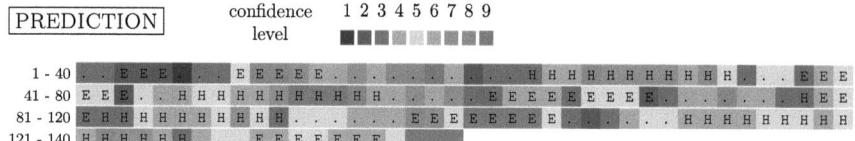

1yln (A1)

helix [H,G,I]		strand [E,B]		coil [C,S,T]		$Q_3^{(loose)}$	$R_3^{(loose)}$
Sens$_3$	Spec$_3$	Sens$_3$	Spec$_3$	Sens$_3$	Spec$_3$		
0.923	0.571	0.724	0.840	0.775	0.775	**0.766**	**0.623**

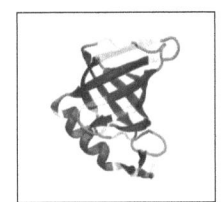

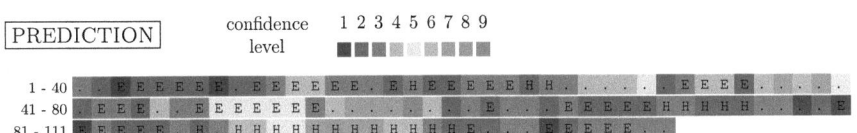

2hu7 (A1)

helix [H,G,I]		strand [E,B]		coil [C,S,T]		$Q_3^{(\text{loose})}$	$R_3^{(\text{loose})}$
Sens$_3$	Spec$_3$	Sens$_3$	Spec$_3$	Sens$_3$	Spec$_3$		
0.706	0.600	0.817	0.865	0.784	0.740	**0.799**	**0.627**

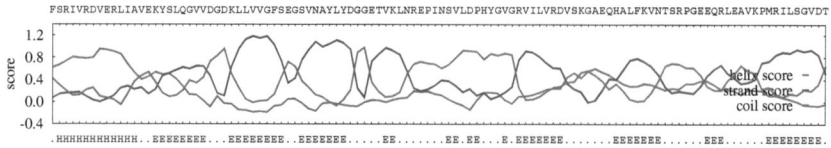

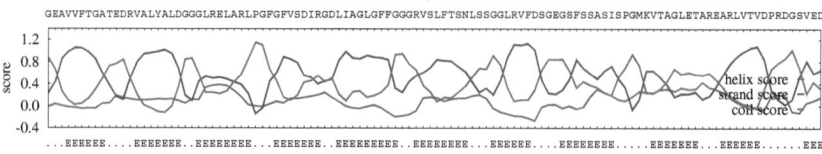

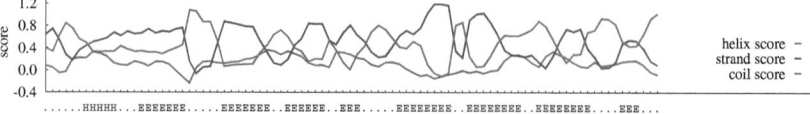

| PREDICTION | confidence level | 1 2 3 4 5 6 7 8 9 |

1 - 40	H H H H H H H H H H H . . H H E E E E E . . . E E E E E E E . . E E E E
41 - 80	E E E E . . E E E E E E . . . . E E E E E E E H H . . H H
81 - 120	H E E E E E E . . . . . . E E E E . . . E E E E E E E E . . E E E E E
121 - 160	E . . E E E E E E . . . E E E E . . . . . . E E E E E . E E E E E E E
161 - 200	E E . . . E E E E E . . . E E E . . . . E E E E E E . . . E E E E . E E
201 - 240	E . . . . . . E E E E E . . . . E E E E E . . . . . E E E E E E E E E E
241 - 280	E E E E E . . . E E E E E . . . E E E E . E E E E . . . . . . E E E E E
281 - 313	E E E . . E E E E E . . . . E E E E . . . . . . E E . .

128

2j8k (A1)

helix [H,G,I]		strand [E,B]		coil [C,S,T]		$Q_3^{(loose)}$	$R_3^{(loose)}$
Sens$_3$	Spec$_3$	Sens$_3$	Spec$_3$	Sens$_3$	Spec$_3$		
1.000	0.529	–	0.677	0.830	0.926	**0.811**	**0.566**

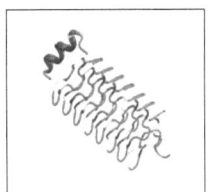

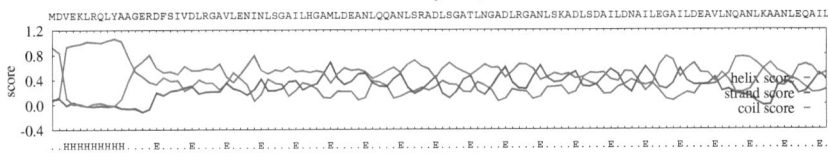

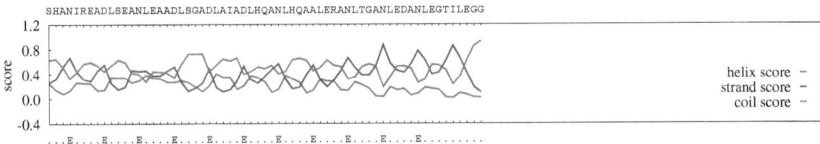

PREDICTION   confidence level   1 2 3 4 5 6 7 8 9

Chapter 7
Conclusions and outlook

The lengthy development that lead to the secondary structure predictor named SPARROW carried a few significant lessons along with it.

First of all, it is clear that a standard representation (see chapter 4) of amino acid sequences seriously limits the performance achievable using straightforward methods like the one based on scoring functions presented here. Clearly enough, treating amino acids as colourless objects with absolute disregard for their physical nature, precludes access to fundamental correlations between primary and secondary structure. The use of more evolved representations like sequence profiles (see chapter 5) is therefore necessary to achieve a level of accuracy comparable to that of powerful predictors such as PSIPRED. Essential to build a competitive scoring function-based predictor is also the inclusion of second order sequence-structure correlations (see quadratic scoring functions in section 4.2). These provide a means to keep up with the parametrical strength of secondary structure predictors based on other machine-learning methods. Refinements like those involving statistical reweighting of different data samples (see section 4.1.1) or learning data set manipulations (see section 4.1.2) on the other hand seem to have mostly a scarce influence on the predictor's performance. Quite relevant are instead enhancements like those deriving from the introduction of the super scoring function (see section 5.2) and the final neural network (see section 5.2.4.4) stages. Devised to effectively combine independent single-state likelihood functions, these make up for the structure-structure correlations missing in the bare sequence-structure correlation machinery.

It is worth to spend one more comment regarding the decoupled treatment of the secondary structure motifs. While this could be considered a drawback as it renders a method based on it uncapable of grasping certain inter-motif correlations, this singular aspect of the developed scoring function-based predictor could present advantages too. It allows in fact to easily focus the attention on one specific secondary structure motif, so that, if required, a more aimed prediction can be performed.

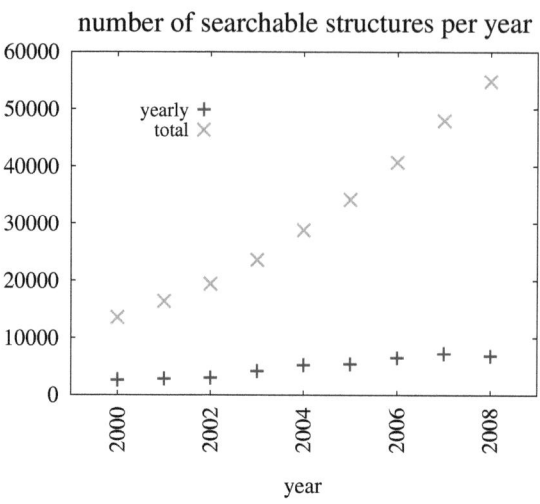

Figure 7.1: Growth of the PDB database [98] in the years from 2000 to 2008.

A modern predictor

The learning method SPARROW relies on, can probably be listed together with those so-called "brute force" methods which have arisen in recent times in response to the outstanding growth of both hardware and database resources. Such growth has made it possible to use models based on a considerable number of parameters, with no risk of overfitting them. Of course, it never ceases to be desirable to have a method capable of maintaining a model's generalization capability independently from the quantity of data available. A way to possibly reduce the thirst of techniques like the one described here could be to restrict the parameter space to a selection of important descriptive features, therewith inhibiting the influence of less relevant information contained in the data sets and thus reducing the number of features necessary to adequately represent those data sets. While this thirst may seem a weakness though, it is important to keep in mind that the amount of available structural data is steadily growing, and possibly at a higher rate each year (see figure 7.1), providing the bioinformatics community with more unbiased data to instruct their prediction machineries with. At the same time, DNA sequencing is constantly improving the quality of sequence profiles [142], the primary input of most secondary structure prediction programs. If such arguments do not negate the value of resource economization, at the very least they de-emphasize it, prompting the structure prediction quest in a more forward-oriented stance. Methods based on neural networks may well remain the most convenient when there is limited data availability but are no longer necessarily the ones of choice in the general case.

It is fair to say in conclusion that the secondary structure predictor called SPARROW, developed in this project, is a valuable tool and can be filed together with the more emblazoned ones existing to date. It can be used as a stand alone or together with other predictors to get an keener insight into the most likely secondary structure organization of a protein. Due to the differences detected in their expertise (see arguments on page 109), it is furthermore not to be excluded that by combining SPARROW with other predictors such as PSIPRED, in the frame of a consensus method, even better results may be achieved.

Future developments

One of the questions left open in part II of the report concerns the true potential residing in higher degrees of secondary structure definition. As established in section 4.1.4, the naive approach followed, barely aids or even harms the prediction performance of the three states, helix, strand and coil. As pointed out in the same section however, there exist techniques, such as those based on genetic algorithms for instance, that may enable an efficient inspection of the reweighting factors and thus unleash the potential of higher degrees of secondary structure definition. The latter could be even better exploited in the super scoring function frame and/or in the final neural network learning stage.

Another possibility currently explored by Dawid Rasinski consists in replacing several single-state likelihood functions with a single *vectorial* gauge function, optimized to indicate pre-assigned points, representing the secondary structure motifs in a multi-dimensional space. It seems this kind of approach may have some advantages and produce even more powerful predictors than the one based on scalar scoring functions presented here.

Part IV
appendices

Appendix A
Single-state prediction accuracy

In the following pages are collected the plots reporting the window length dependence of the single-state accuracy indices.

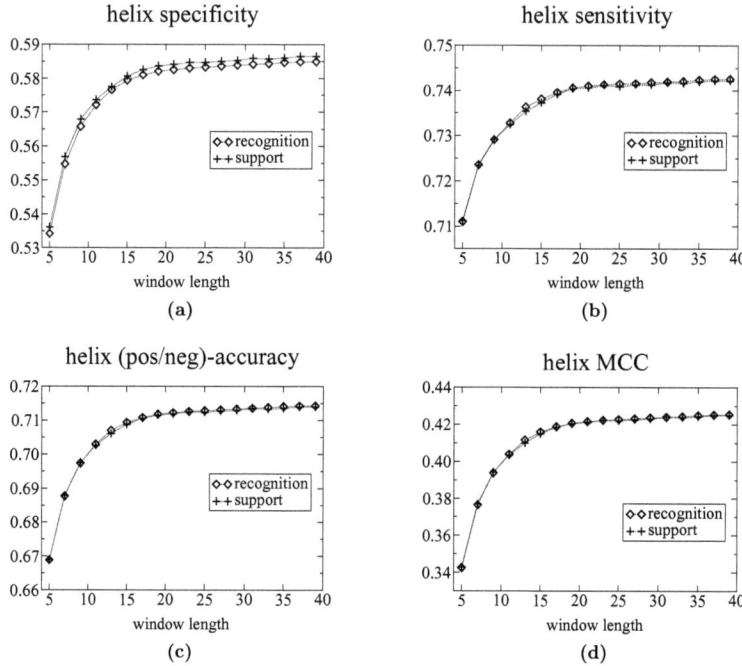

Figure A.1: Recognition and supporting test prediction quality indices for helix. The supporting data set consisted in sequence samples subtracted from the learning set amounting to roughly 10% of the latter. The average values were computed using (2.3.1); the deviations are not reported. The functions used were linear and optimized using non-reweighted dual statistics. Only the central secondary structure key-position was used.

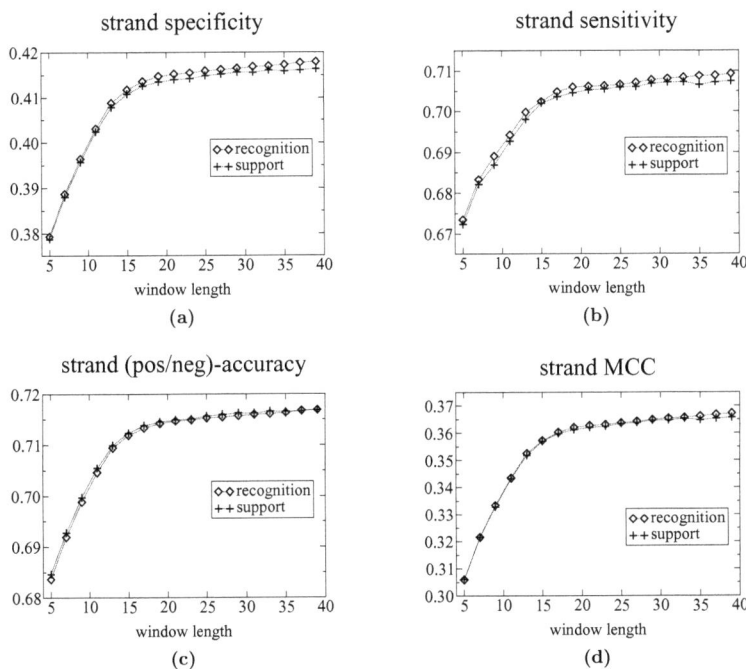

Figure A.2: Recognition and supporting test prediction quality indices for strand. The supporting data set consisted in sequence samples subtracted from the learning set amounting to roughly 10% of the latter. The average values were computed using (2.3.1); the deviations are not reported. The functions used were linear and optimized using non-reweighted dual statistics. Only the central secondary structure key-position was used.

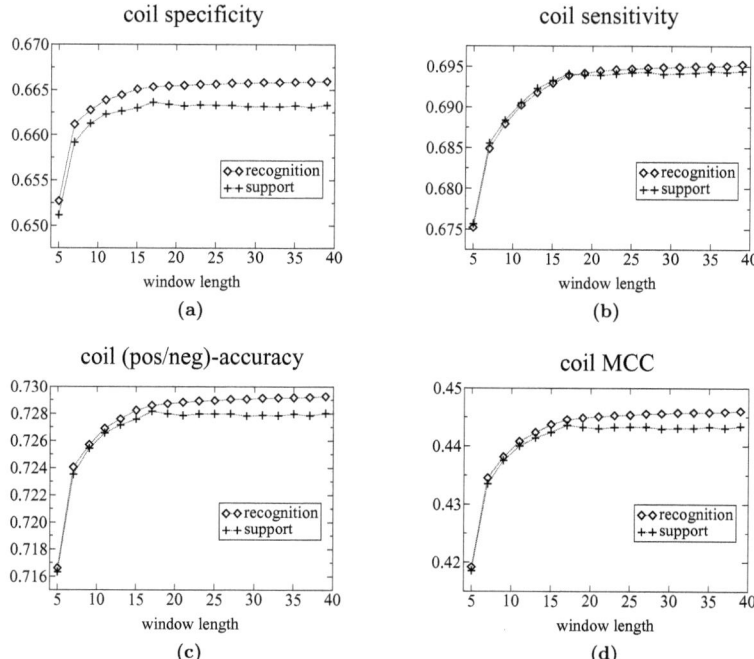

Figure A.3: Recognition and supporting test prediction quality indices for coil. The supporting data set consisted in sequence samples subtracted from the learning set amounting to roughly 10% of the latter. The average values were computed using (2.3.1); the deviations are not reported. The functions used were linear, based on a sequence window of 19 residues and optimized using non-reweighted dual statistics. Only the central secondary structure key-position was used. Note that the size of the sequence window affects the coil quality indices to a much smaller extent than the helix and strand quality indices.

Table A.1: Supporting test prediction quality indices for helix strand and coil as a function of the sequence window lengths. The values for recognition are comparable. Only the results for sequence windows with up to 19 amino acids are shown.

	L	specificity	sensitivity	pos/neg accuracy	MCC
helix	5	0.536 ± 0.002	0.711 ± 0.002	0.670 ± 0.001	0.343 ± 0.002
	7	0.557 ± 0.002	0.723 ± 0.002	0.688 ± 0.001	0.377 ± 0.002
	9	0.568 ± 0.002	0.729 ± 0.002	0.698 ± 0.001	0.394 ± 0.001
	11	0.574 ± 0.002	0.732 ± 0.002	0.703 ± 0.001	0.403 ± 0.002
	13	0.577 ± 0.002	0.735 ± 0.002	0.706 ± 0.001	0.410 ± 0.001
	15	0.581 ± 0.002	0.737 ± 0.002	0.709 ± 0.001	0.415 ± 0.001
	17	0.583 ± 0.002	0.739 ± 0.002	0.710 ± 0.001	0.418 ± 0.002
	19	0.584 ± 0.002	0.740 ± 0.002	0.711 ± 0.001	0.420 ± 0.002
	...			...	
strand	5	0.379 ± 0.002	0.672 ± 0.002	0.685 ± 0.001	0.306 ± 0.002
	7	0.388 ± 0.002	0.682 ± 0.002	0.693 ± 0.001	0.321 ± 0.002
	9	0.396 ± 0.002	0.687 ± 0.003	0.700 ± 0.001	0.333 ± 0.002
	11	0.403 ± 0.002	0.693 ± 0.003	0.705 ± 0.001	0.343 ± 0.002
	13	0.408 ± 0.002	0.698 ± 0.003	0.710 ± 0.001	0.352 ± 0.002
	15	0.411 ± 0.002	0.702 ± 0.002	0.712 ± 0.001	0.357 ± 0.002
	17	0.413 ± 0.002	0.704 ± 0.003	0.714 ± 0.001	0.360 ± 0.002
	19	0.414 ± 0.002	0.705 ± 0.002	0.714 ± 0.001	0.361 ± 0.002
	...			...	
coil	5	0.651 ± 0.001	0.676 ± 0.002	0.716 ± 0.001	0.418 ± 0.001
	7	0.659 ± 0.001	0.685 ± 0.002	0.723 ± 0.001	0.433 ± 0.002
	9	0.661 ± 0.002	0.688 ± 0.002	0.725 ± 0.001	0.437 ± 0.001
	11	0.662 ± 0.001	0.690 ± 0.002	0.727 ± 0.001	0.440 ± 0.002
	13	0.663 ± 0.002	0.692 ± 0.002	0.727 ± 0.001	0.441 ± 0.002
	15	0.663 ± 0.002	0.693 ± 0.002	0.728 ± 0.001	0.442 ± 0.002
	17	0.664 ± 0.002	0.694 ± 0.002	0.728 ± 0.001	0.444 ± 0.002
	19	0.663 ± 0.001	0.694 ± 0.002	0.728 ± 0.001	0.443 ± 0.001
	...			...	

Appendix B
SPARROW prediction data

Table B.1 contains the results, complete of all quality indices, obtained predicting the secondary structure of the domains in ASTRAL40 – release 1.73 with up to 40% sequence identity with those in ASTRAL40 – release 1.71.

Table B.1: SPARROW's overall and partial multi-choice prediction accuracy for helix, strand and coil classes, on the domains of ASTRAL40 – release 1.73 with up to 40% sequence identity with the domains in ASTRAL40 – release 1.71.

	DSSP	aligned sequences		non-aligned sequences	
		Sens_3	Spec_3	Sens_3	Spec_3
helix	[H,G,I]	0.8475	0.8704	0.7656	0.8300
strand	[E,B]	0.7412	0.7992	0.5267	0.4540
coil	[C,S,T]	0.8104	0.7620	0.7877	0.7614
$Q_3^{(\text{loose})}$		**0.8091**		**0.7498**	
$R_3^{(\text{loose})}$		**0.6757**		**0.4935**	

The following pages feature the partial and cumulative plots for the accuracy indices which were not reported in the body of the report. The domains are sorted with respect to sequence identity, sequence similarity (percentage of residues positively contributing to the BLAST score), BLAST score and p-value.

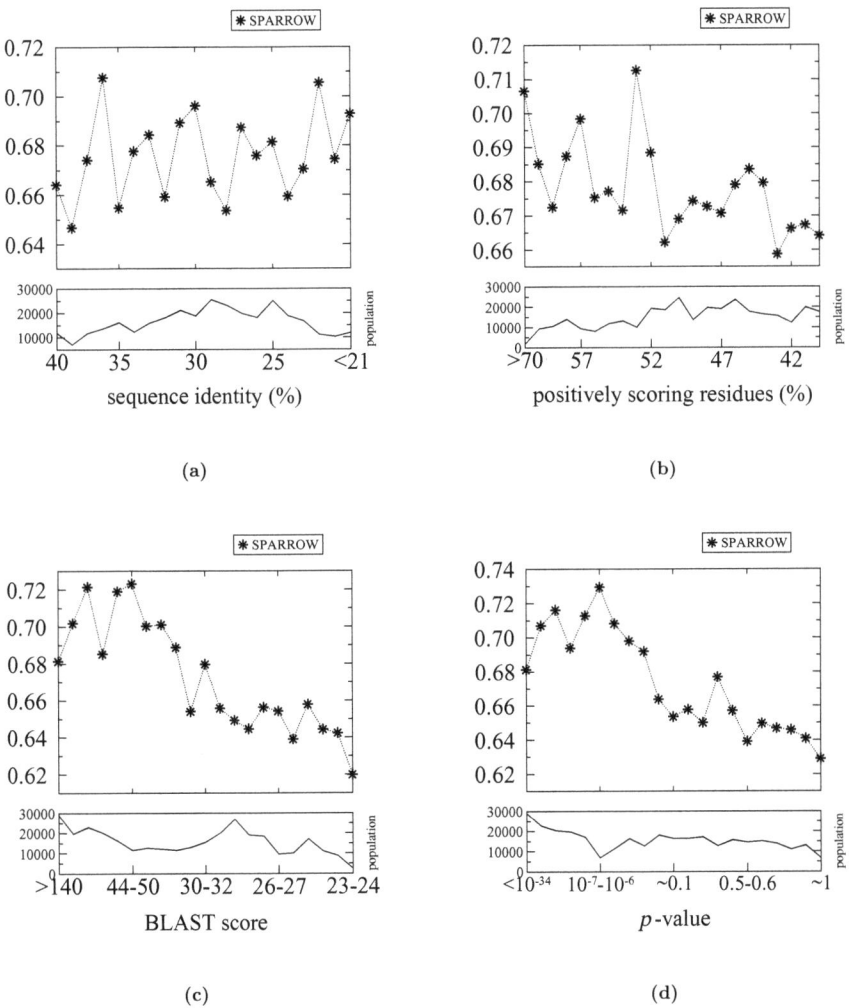

Figure B.1: R_3 correlation coefficient achieved by SPARROW on the domains of ASTRAL40 – release 1.73, sorted respectively (a) by percentage of sequence identity, (b) by percentage of residues that positively contributed to the BLAST alignment score, (c) by BLAST alignment score and (d) by p-value.

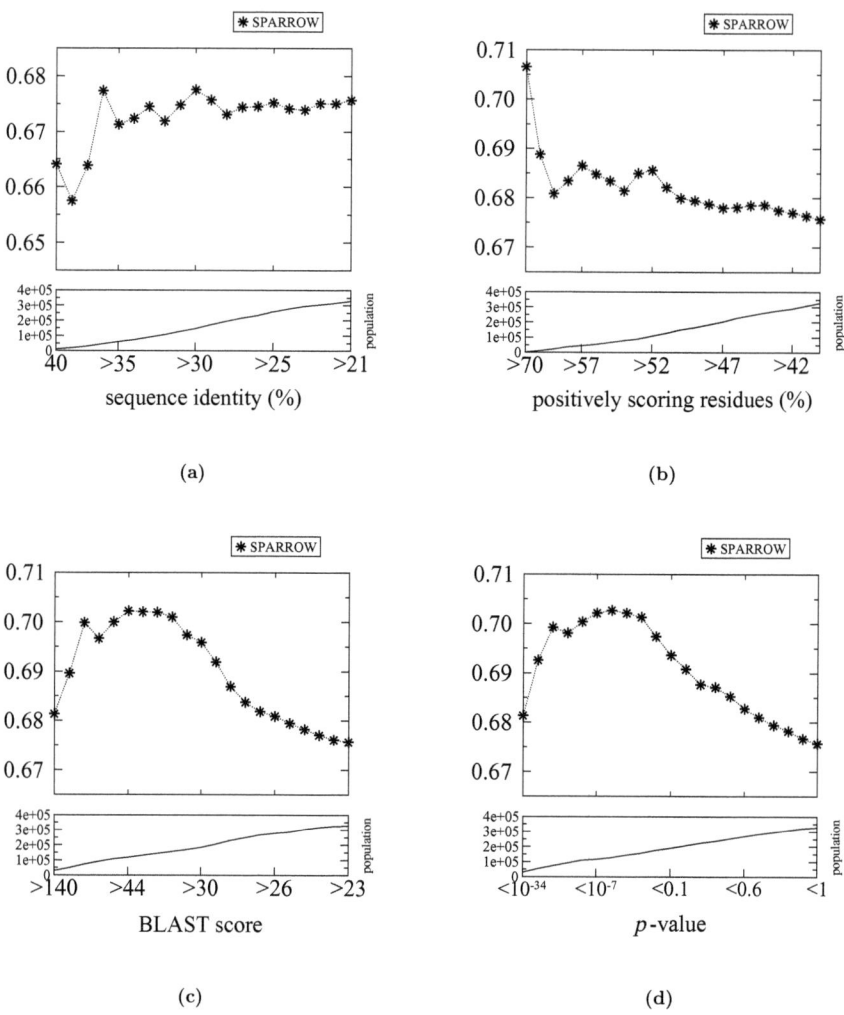

Figure B.2: R_3 correlation coefficient achieved by SPARROW on an increasingly greater portion of the domains of ASTRAL40 – release 1.73, obtained by progressively adding chains with respectively (a) lower percentage of sequence identity, (b) lower percentage of residues that positively contributed to the BLAST alignment score, (c) lower BLAST alignment score and (d) higher p-value.

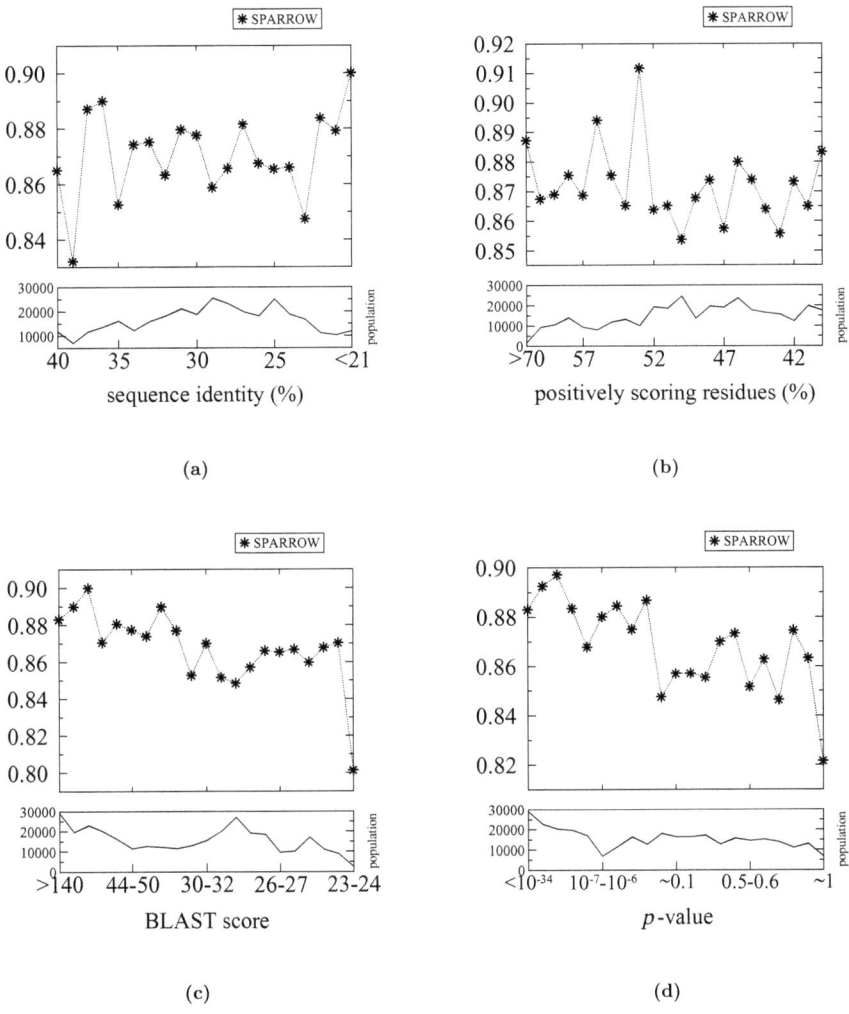

Figure B.3: Helix-specificity of SPARROW on the domains of AstRAL40 – release 1.73, sorted respectively (a) by percentage of sequence identity, (b) by percentage of residues that positively contributed to the BLAST alignment score, (c) by BLAST alignment score and (d) by p-value.

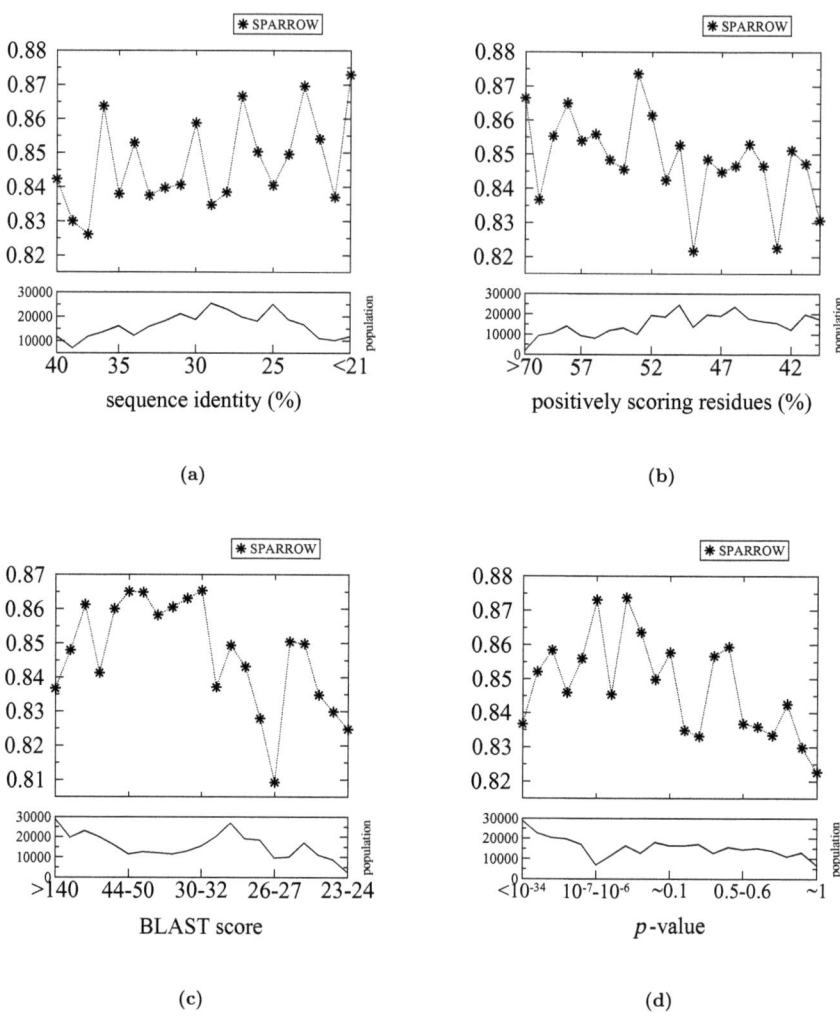

Figure B.4: Helix-sensitivity of SPARROW on the domains of ASTRAL40 – release 1.73, sorted respectively (a) by percentage of sequence identity, (b) by percentage of residues that positively contributed to the BLAST alignment score, (c) by BLAST alignment score and (d) by p-value.

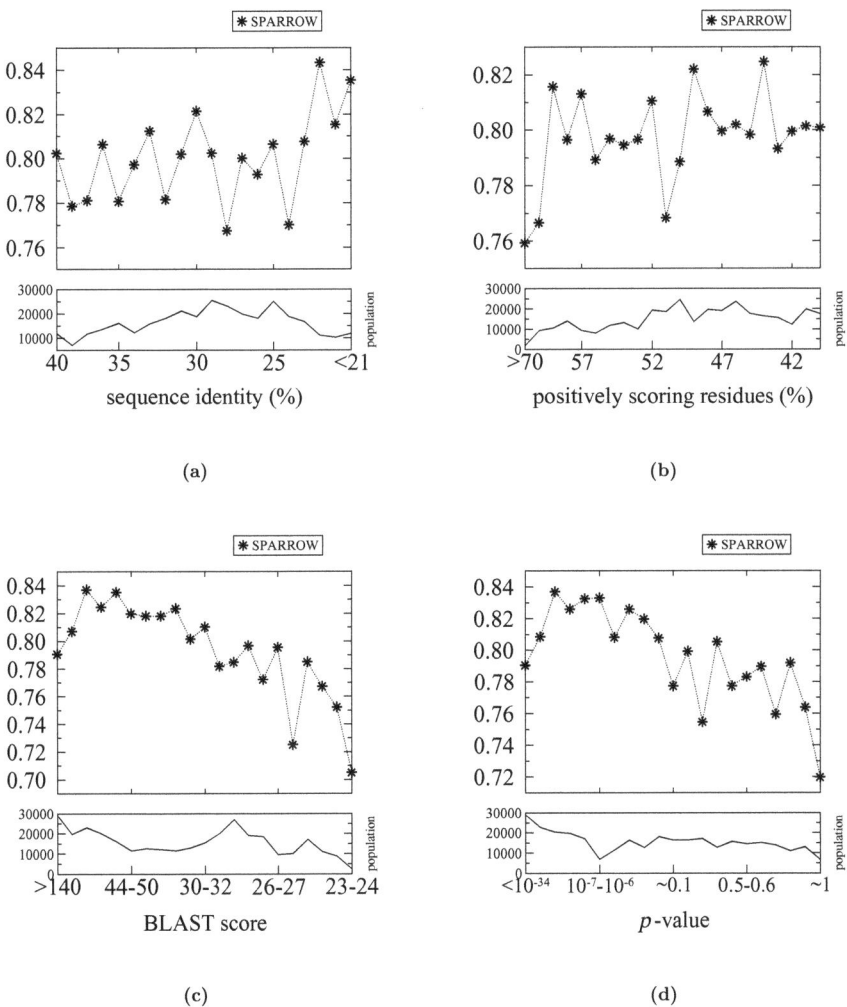

Figure B.5: Strand-specificity of SPARROW on the domains of ASTRAL40 – release 1.73, sorted respectively (a) by percentage of sequence identity, (b) by percentage of residues that positively contributed to the BLAST alignment score, (c) by BLAST alignment score and (d) by p-value.

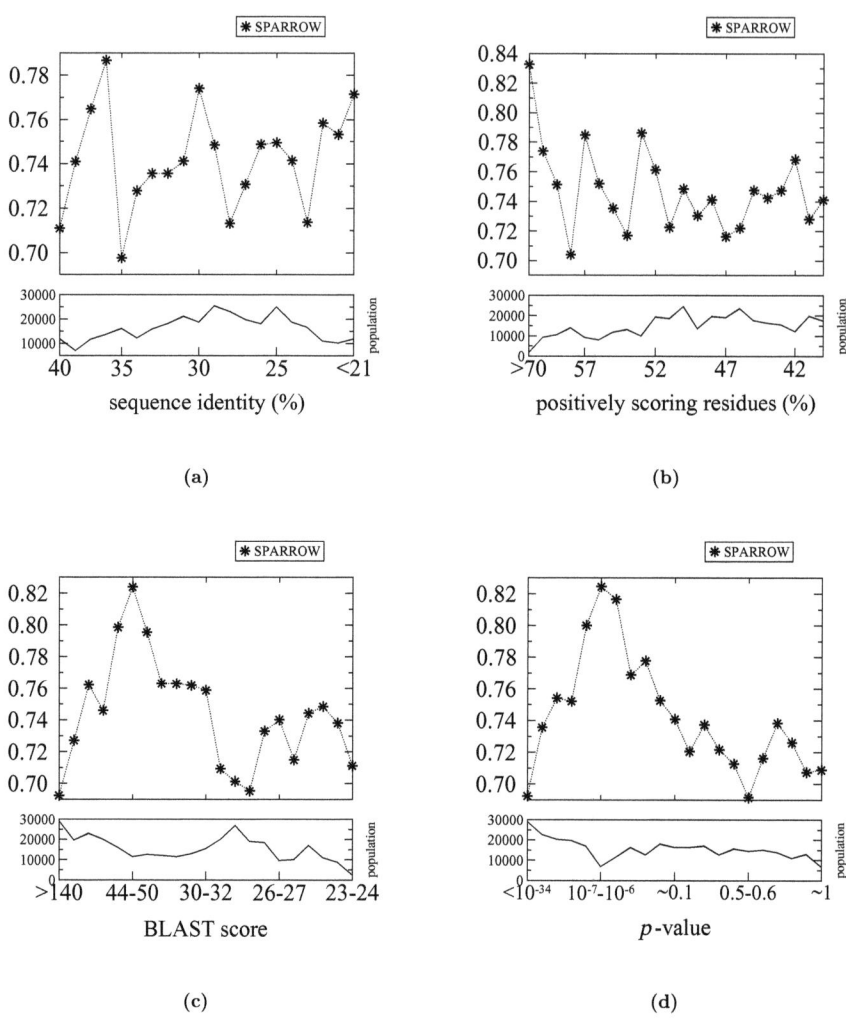

Figure B.6: Strand-sensitivity of SPARROW on the domains of ASTRAL40 – release 1.73, sorted respectively (a) by percentage of sequence identity, (b) by percentage of residues that positively contributed to the BLAST alignment score, (c) by BLAST alignment score and (d) by p-value.

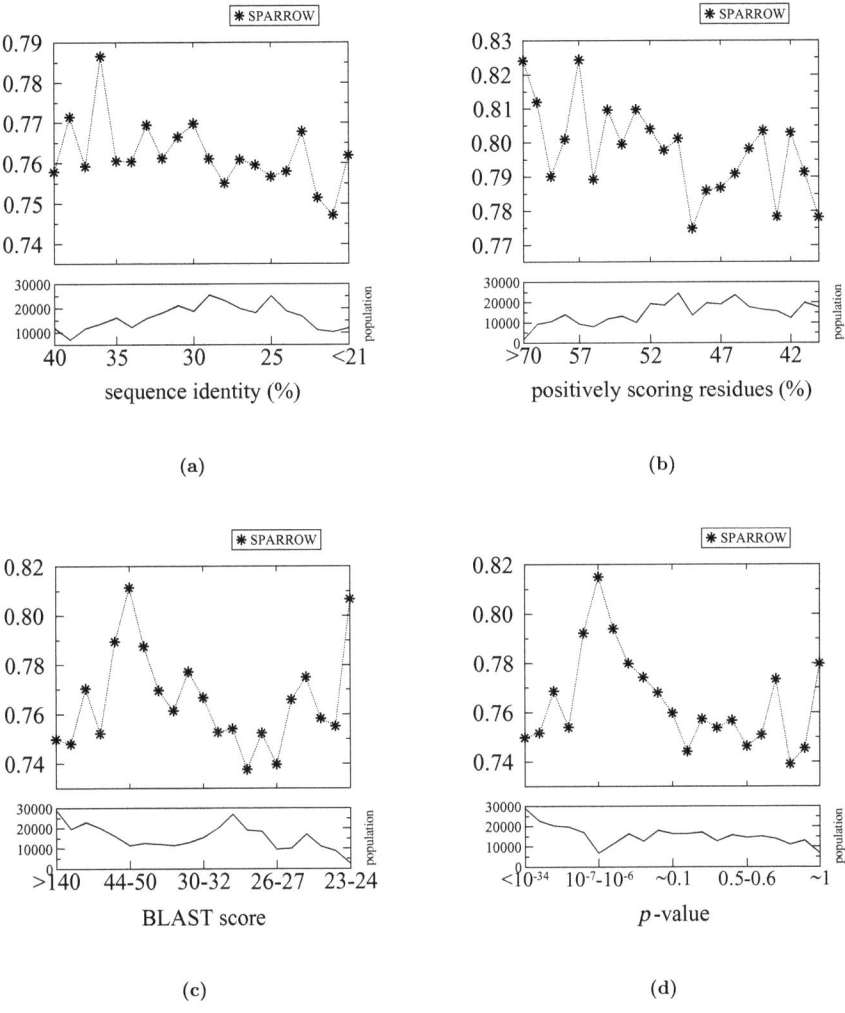

Figure B.7: Coil-specificity of SPARROW on the domains of ASTRAL40 – release 1.73, sorted respectively (a) by percentage of sequence identity, (b) by percentage of residues that positively contributed to the BLAST alignment score, (c) by BLAST alignment score and (d) by p-value.

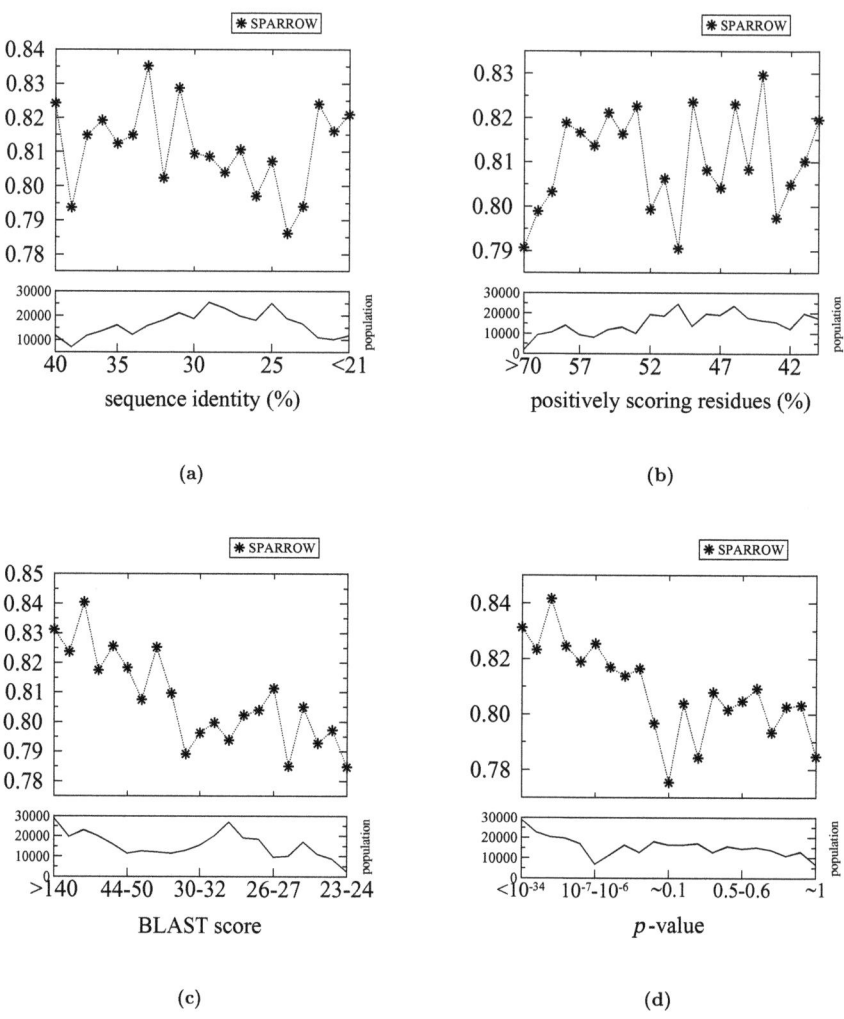

Figure B.8: Coil-sensitivity of SPARROW on the domains of ASTRAL40 – release 1.73, sorted respectively (a) by percentage of sequence identity, (b) by percentage of residues that positively contributed to the BLAST alignment score, (c) by BLAST alignment score and (d) by p-value.

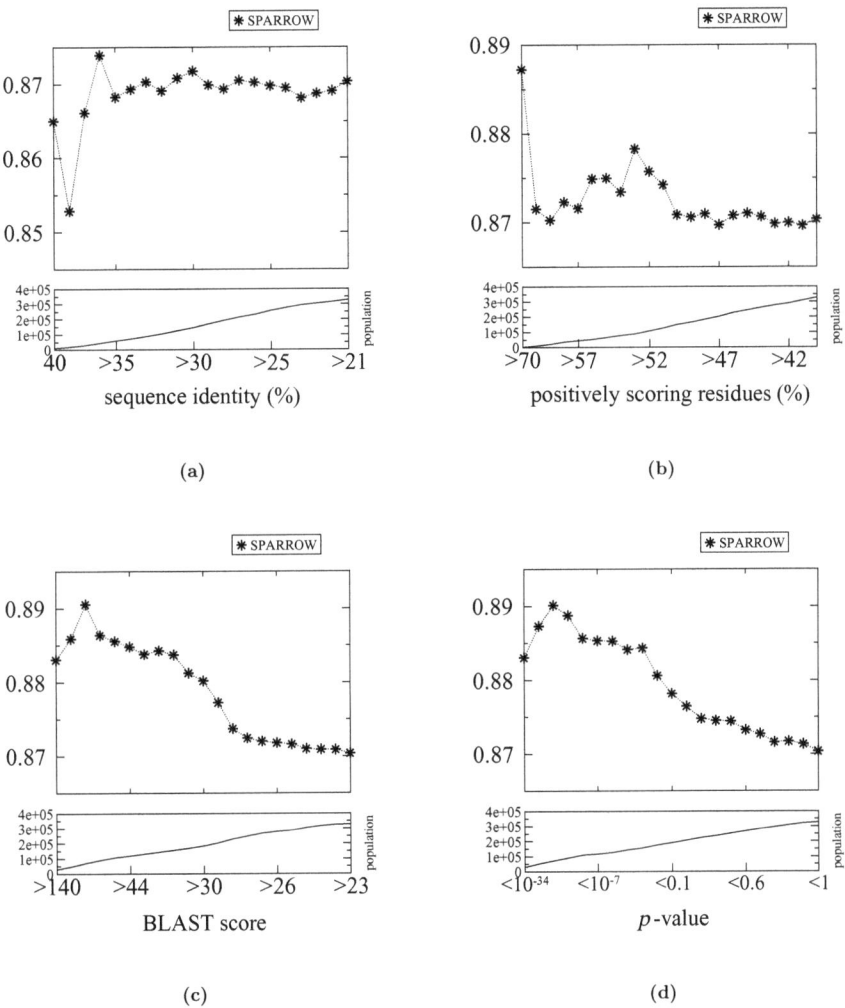

Figure B.9: Helix-specificity of SPARROW on an increasingly greater portion of the domains of ASTRAL40 – release 1.73, obtained by progressively adding chains with respectively (a) lower percentage of sequence identity, (b) lower percentage of residues that positively contributed to the BLAST alignment score, (c) lower BLAST alignment score and (d) higher p-value.

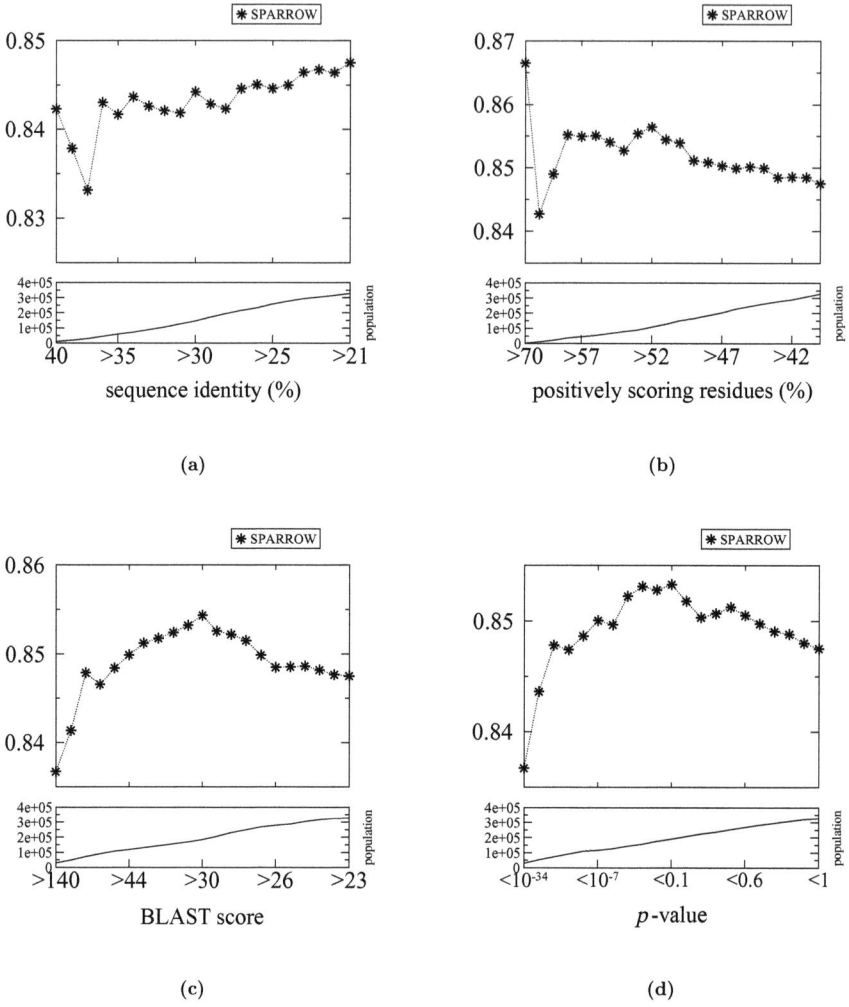

Figure B.10: Helix-sensitivity of SPARROW on an increasingly greater portion of the domains of ASTRAL40 – release 1.73, obtained by progressively adding chains with respectively (a) lower percentage of sequence identity, (b) lower percentage of residues that positively contributed to the BLAST alignment score, (c) lower BLAST alignment score and (d) higher p-value.

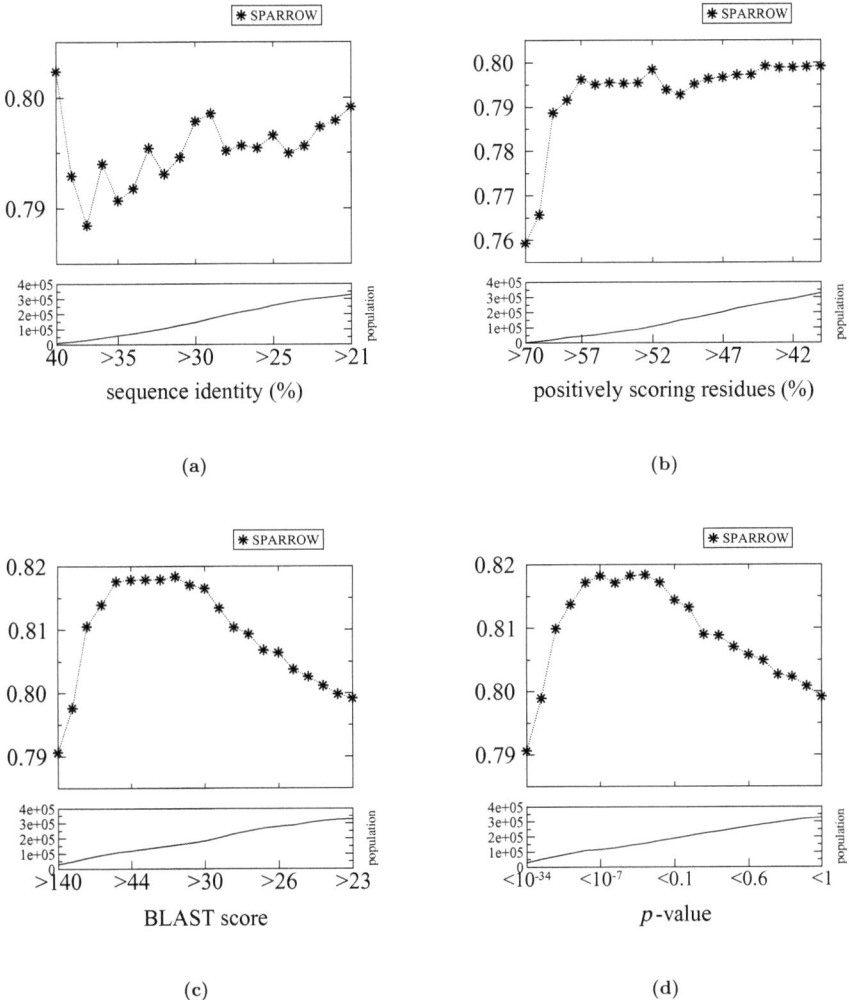

Figure B.11: Strand-specificity of SPARROW on an increasingly greater portion of the domains of ASTRAL40 – release 1.73, obtained by progressively adding chains with respectively (a) lower percentage of sequence identity, (b) lower percentage of residues that positively contributed to the BLAST alignment score, (c) lower BLAST alignment score and (d) higher p-value.

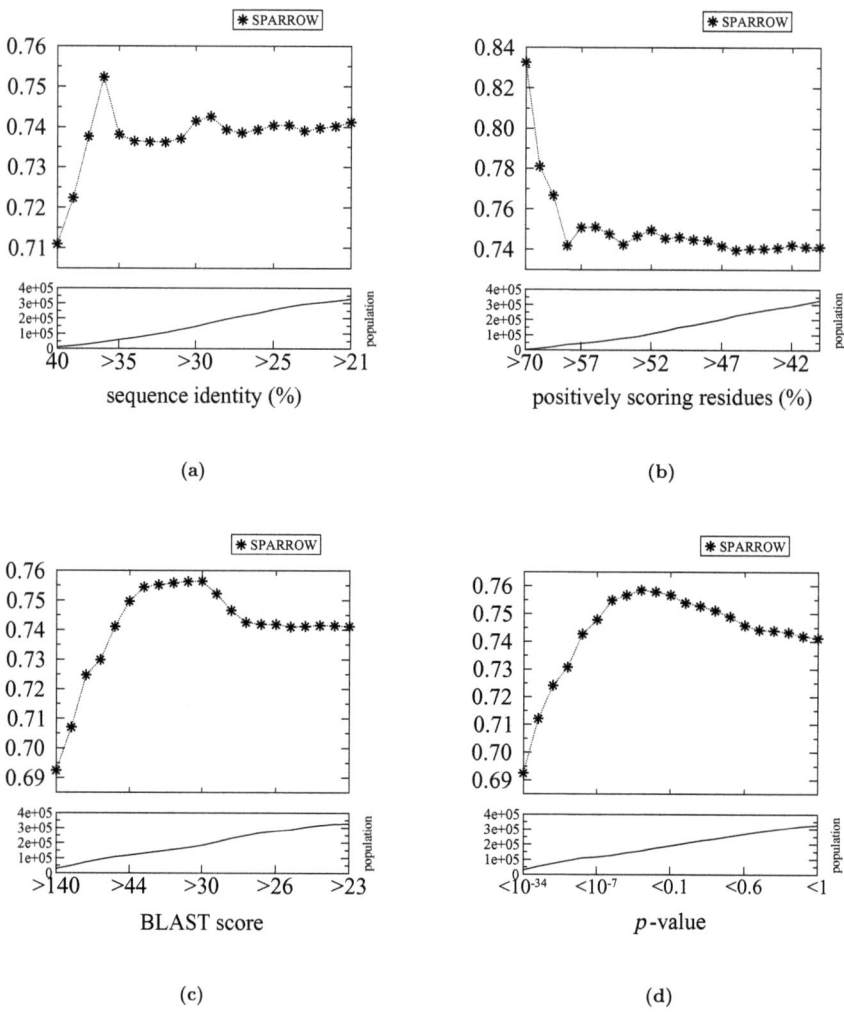

Figure B.12: Strand-sensitivity of SPARROW on an increasingly greater portion of the domains of ASTRAL40 – release 1.73, obtained by progressively adding chains with respectively (a) lower percentage of sequence identity, (b) lower percentage of residues that positively contributed to the BLAST alignment score, (c) lower BLAST alignment score and (d) higher p-value.

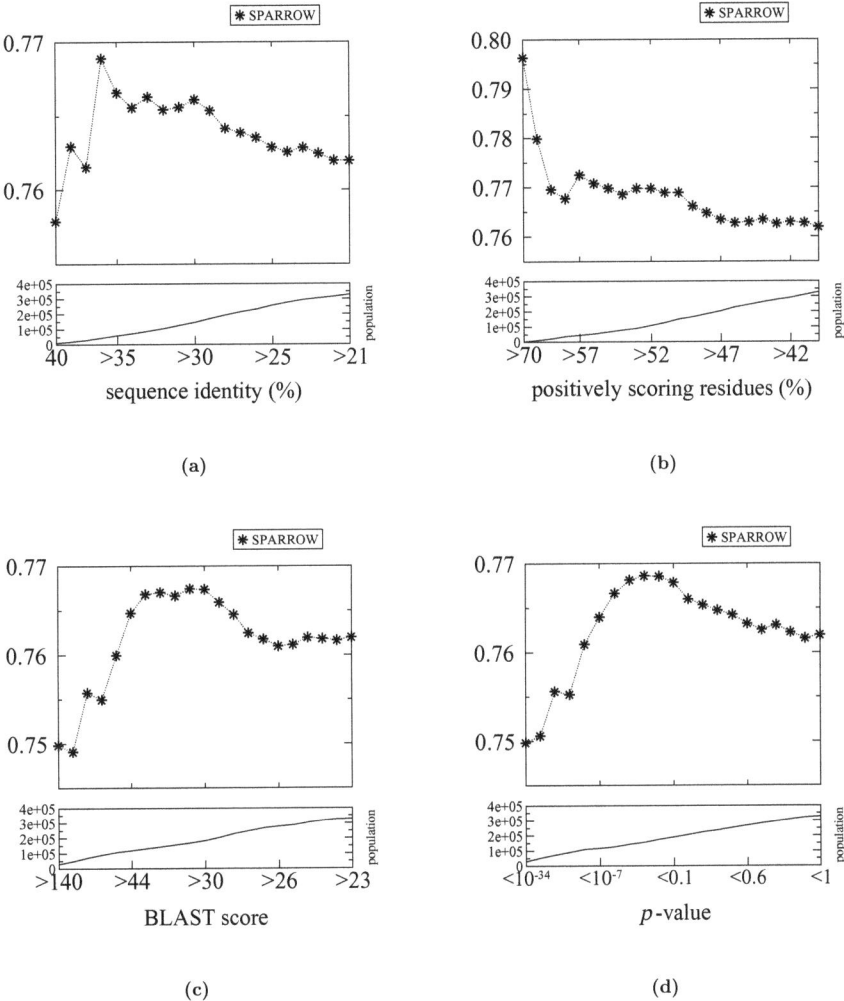

Figure B.13: Coil-specificity of SPARROW on an increasingly greater portion of the domains of ASTRAL40 – release 1.73, obtained by progressively adding chains with respectively (a) lower percentage of sequence identity, (b) lower percentage of residues that positively contributed to the BLAST alignment score, (c) lower BLAST alignment score and (d) higher p-value.

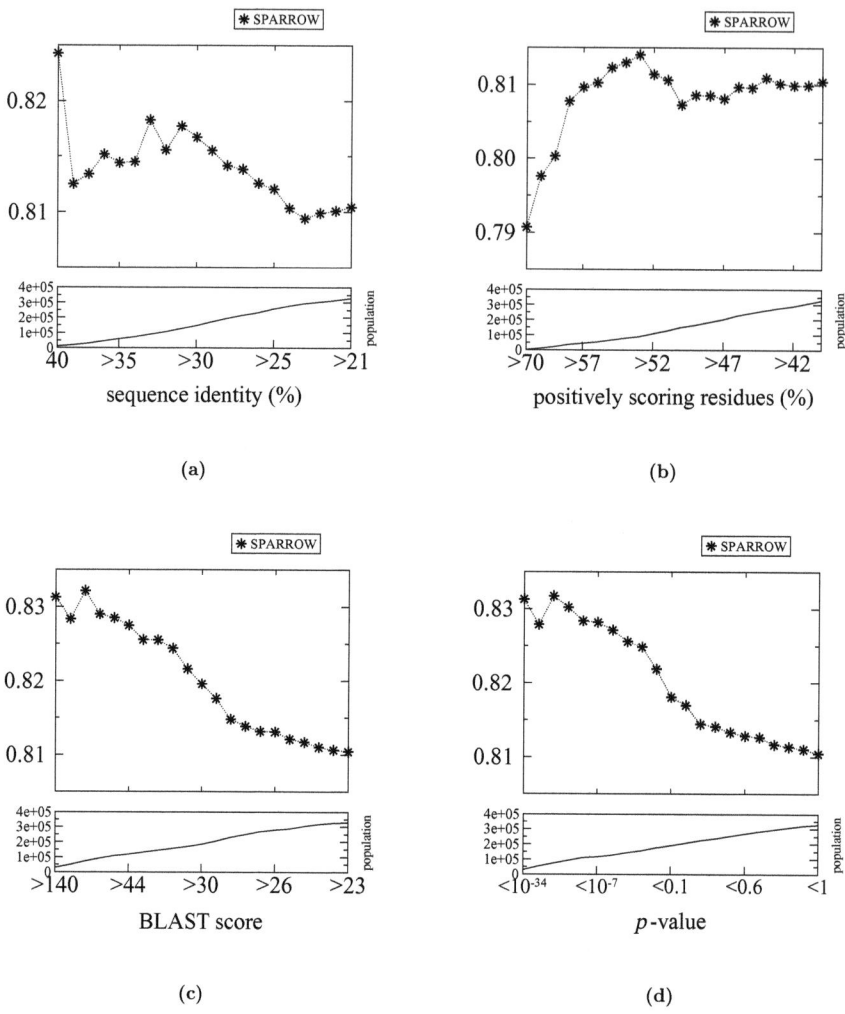

Figure B.14: Coil-sensitivity of SPARROW on an increasingly greater portion of the domains of ASTRAL40 – release 1.73, obtained by progressively adding chains with respectively (a) lower percentage of sequence identity, (b) lower percentage of residues that positively contributed to the BLAST alignment score, (c) lower BLAST alignment score and (d) higher p-value.

Summary

Proteins are essential constituents of all living organisms. Most proteins fold into unique structures determined by the sequence of amino acids composing them. In three-dimensional protein structures, regularly recurrent local structural motifs like α-helices and β-sheets can often be identified. Such local arrangements are collectively called *secondary structure*, while the way in which a polypeptide chain finally folds in the three-dimensional space is called *tertiary structure*.

Since the structure of a protein plays such a central role for its function within living organisms, it is subject of great interest. Experimental techniques have been developed to investigate it, but are relatively expensive and time consuming. As a consequence, the need is increasingly felt for theoretical structure prediction methods.

Predicting the structure of proteins is a very difficult task. Several protein structure prediction methods exist to date. A considerable aid is provided to these methods if a prediction of the secondary structure of the protein is available beforehand. Reliable secondary structure information can be employed, for example, to build safe starting cores in fold simulation programs or structural constraints in protein threading and homology modelling searches.

Aim of this project was to develop a new tool to predict protein secondary structure from the amino acid sequence.

The method employed is of statistical nature and relies on existing protein data. The primary structure was input in the form of PSI-BLAST profiles. The secondary structure information used to instruct the program was extracted from atomic coordinates using DSSP.

Though secondary structure prediction is a three-choice classification problem, the approach adopted was to consider all motifs separately and reduce it to the sum of three single-choice classification problems. This is done by grouping all the motifs that are not the one of interest into a single class and having the program learn a set of rules to respectively sort the one of interest out. It turns out that each set of rules won through this procedure provides a means of measuring the *likelihood* that some amino acid sequence be associated to the corresponding secondary structure motif. Once all the needed sets of this kind are available, a direct comparison of the likelihoods attainable from them allows to solve the three-choice classification problem. In more recent developments a neural network was deployed to make the best sense out of the likelihood scores and perform the three-choice prediction based on them.

In a ten-fold cross-validation based on the release 1.71 of Astral40, the current version of the program achieved an average accuracy of about 82% in predicting which secondary structure motif among helix, strand and coil (none), a residue adopts. Other prediction tests carried out on the release 1.73 of Astral40 show that the developed secondary structure predictor can compete with a celebrated predictor like PSIPRED.

Zusammenfassung

Proteine sind wesentliche Bestandteile von allen lebenden Organismen. Die meisten Proteine falten sich in einzigartigen Strukturen, die von der Sequenz der bildenden Aminosäuren bestimmt werden. In dreidimensionalen Proteinstrukturen treten regelmäßig wiederkehrende lokale strukturelle Motive wie α-Helices und β-Strands oft auf. Solche lokalen Anordnungen werden in ihrer Gesamtheit als *Sekundärstruktur* bezeichnet, während die Art und Weise, in der sich eine Polypeptidkette schließlich im dreidimensionalen Raum faltet, *Tertiärstruktur* genannt wird.

Da die Struktur eines Proteins eine so zentrale Rolle für die Funktion des Proteins selbst in lebenden Organismen spielt, ist sie von großem Interesse. Experimentelle Techniken wurden entwickelt, um sie zu untersuchen. Diese sind aber relativ teuer und zeitaufwendig. Daher ist eine wachsende Notwendigkeit für theoretische Methoden zur Strukturvorhersage zu spüren.

Proteinstrukturvorhersage ist eine sehr schwierige Aufgabe. Mehrere Methoden sind bis heute entwickelt worden. Eine wichtige Beihilfe für diese Methoden entsteht, wenn eine Sekundärstrukturvorhersage vorhanden ist. Zuverlässige Sekundärstrukturinformation kann zum Beispiel zum Aufbau von sicheren Kernen in Fold-Simulation Programmen oder strukturellen Beschränkungen in Protein Threading und Homology-Modelling-Suchen eingesetzt werden.

Ziel dieses Projektes war es, ein neues Instrument für die Proteinsekundärstrukturvorhersage aus der Sequenz der Aminosäuren zu entwickeln.

Die angewendete Methode ist von statistischer Natur und basiert auf bestehenden Proteindaten. Die Primärstruktur wurde in Form von PSI-BLAST-Profilen betrachtet. Die zum Lernen eingesetzte Sekundärstruktur wurde aus den Atomkoordinaten durch DSSP hergeleitet.

Obwohl Sekundärstrukturvorhersage ein Dreiklassenproblem ist, war der angenommene Ansatz alle Motive einzeln zu betrachten und es in die Summe von drei Zweiklassenprobleme umzuwandeln. Dies ist durch Gruppieren der jeweils nicht betrachteten Motive zu einer einzigen Klasse und das Lernen von Regeln, um das betrachtete Motiv auszusortieren, erfolgt. Es stellt sich heraus, dass jeder durch ein solches Verfahren gewonnene Regelsatz ein Mittel zur Messung der Wahrscheinlichkeit, dass eine Aminosäuresequenz zu der entsprechenden Sekundärstruktur führt, darstellt. Sobald alle benötigten Regelsätze dieser Art zur Verfügung stehen, ermöglicht ein direkter Vergleich der durch diese bestimmten Wahrscheinlichkeiten, das Dreiklassenproblem zu lösen. In der letzten Entwicklung wurde ein neuronales Netz eingesetzt, um die Wahrscheinlichkeiten zu verbessern und die Dreifachklassifizierung durchzuführen.

In einer auf Version 1.71 von Astral40 basierenden zehnfachen Kreuzvalidierung erreichte die aktuelle Version des Programms eine durchschnittliche genauigkeit von ungefähr 82% in der Vorhersage welches von den Motiven, Helix, Strand und Coil (kein Motiv), eine Aminosäure einnimmt. Weitere Vorhersage Tests auf der Version 1.73 von Astral40 zeigen, dass die in diesem Projekt entwickelte Sekundärstrukturvorhersage-Software in der Lage ist, auf dem gleichen Niveau wie ein etabliertes Programm wie PSIPRED zu arbeiten.

Acknowledgments

Part of the work described in this report has been carried out in cooperation with Dawid Rasinski. In particular, the development of the artificial neural network is to be accredited to him in its entirety. To the whole AGKnapp and especially to Tobias Schmidt-Gönner goes my appreciation for several very useful discussions.

I finally wish to express my deepest gratitude to my friend and partner Ana Patricia L Gamiz Hernandez who always supported and comforted me along my whole Ph.D. endeavour.

Bibliography

[1] Erwin Schrödinger, *What Is Life?*, Cambridge University Press, 1944.

[2] Lynn Margulis and Dorion Sagan, *What Is Life?*, Simon & Schuster, 1995.

[3] Steven Levy, *Artificial Life*, Pantheon Books, 1992.

[4] Christopher G. Langton, "Computation at the edge of chaos", *Physica D* **42**, 1990.

[5] Roger Lewin, *Complexity: Life at the Edge of Chaos*, Macmillan Publishing Company, 1992.

[6] M. Mitchell Waldrop, *Complexity: The Emerging Science at the Edge of Order and Chaos*, Simon & Schuster, 1992.

[7] Stuart A. Kauffman, *The Origins of Order: Self-Organization and Selection in Evolution*, Oxford University Press, 1993.

[8] Stuart A. Kauffman, *At Home in the Universe: The Search for Laws of Self-organization and Complexity*, Oxford University Press, 1995.

[9] Anfinsen C. B., Haber E., Sela M. and White F. H., "The kinetics of formation of native ribonuclease during oxidation of the reduced polypeptide chain", *Proc. Nat. Acad. Sci.* **47**:1309–1314, 1961.

[10] Pauling L., Corey R. B. and Branson H. R., "Two hydrogen-bonded helical configurations of the polypeptide chain", *Proc. Natl. Acad. Sci. Wash.* **37**:205–211, 1951.

[11] Pauling L., Corey R. B., "Configurations of polypeptide chains with favored orientations of the polypeptide around single bonds: two pleated sheets", *Proc. Natl. Acad. Sci. Wash.* **37**:729–740, 1951.

[12] Linderstrøm Lang K. U., "Proteins and Enzymes", *Lane Medical Lectures* 6, 1952.

[13] Marqusee S. and Baldwin R. L., "Helix stabilization by $Glu^-\cdots Lys^+$ salt bridges in short peptides of de novo design", *Proc. Nat. Acad. Sci.* **84**:8898–8902, 1987.

[14] Marqusee S., Robbins V. J. and Baldwin R. L., "Unusually stable helix formation in short alanine-based peptides", *Proc. Nat. Acad. Sci.* **86**:5286–5290, 1989.

[15] Oas T. G. and Kim I. S., "A peptide model of a protein folding intermediate", *Nature* **336**:42–48, 1988.

[16] Roder H., Elove G. A. and Englander S. N., "Structural characterization of folding intermediates in cytochrome-c by H-exchange labelling and proton NMR", *Nature* **335**:700–704, 1988.

[17] Udgaonkar J. B. and Baldwin R. L., "NMR evidence for an early framework intermediate on the folding pathway of ribonuclease A", *Nature* **335**:694–699, 1988.

[18] Ramachandran G. N., Ramakrishnan C. and Sasisekharan V., "Stereochemistry of polypeptide chain configurations", *J. Mol. Biol.* **7**:95–99, 1963.

[19] Chandonia J. M., Hon G., Walker N. S., Lo Conte L., Koehl P., Levitt M., Brenner S.E., "The ASTRAL compendium in 2004", *Nucleic Acids Research* **32**:189–192, 2004.

[20] Yue K. and Dill K. A., "Folding proteins with a simple energy function and extensive conformational searching", *Protein Science* **5**(2):254–261, 1996.

[21] Park K., Vendruscolo M., Domany E., "Toward an energy function for the contact map representation of proteins", *Proteins: structure, function and bioinformatics* **40**(2):237–248, 2000.

[22] Vendruscolo M. Domany E., "Protein folding using contact maps", *Vitam. Horm.* **58**:171–212, 2000.

[23] Zhang Y. and Skolnick J., "The protein structure prediction problem could be solved using the current PDB library", *Proc. Natl. Acad. Sci.* **102**:1029–1034, 2005.

[24] Bowie J. U., Luthy R., Eisenberg D., "A method to identify protein sequences that fold into a known three-dimensional structure", *Science* **253**:164–170, 1991.

[25] Burgess A. W., Scheraga H. A., "Assessment of some problems associated with prediction of 3-dimensional structure of a protein from its amino-acid sequence", *Proc. Natl. Acad. Sci.* **72**:1221–1225, 1975.

[26] Jones D. T., Taylor W. R., Thornton J. M., "A new approach to protein fold recognition", *Nature* **358**:86–89, 1992.

[27] Solis A. D. and Rackovsky S., "On the use of secondary structure in protein structure prediction: a bioinformatic analysis", *Polymer* **45**:525–546, 2004.

[28] Colubri A. D. and Rackovsky S., "Prediction of protein structure by simulating coarse-grained folding pathways: a preliminary report", *J. Biomol. Str. & Dyn.* **21**:625–638, 2004.

[29] Fischer D., Eisenberg D., "Protein fold recognition using sequence-derived predictions", *Protein Science* **5**:947–955, 1996.

[30] Rost B., Schneider R., Sander C., "Protein fold recognition by prediction-based threading", *J. Mol. Biol.* **270**:471–480, 1997.

[31] Russell R. B., Copley R. R. and Barton G. J., "Protein fold recognition by mapping predicted secondary structures", *J. Mol. Biol.* **259**:349–365, 1996.

[32] Karchin R., Cline M., Mandel-Gutfreund Y., Karplus K., "Hidden Markov models that use predicted local structure for fold recognition: alphabets of backbone geometry", *Proteins* **51**(4):504–514, 2003.

[33] Sim J., Kim S. Y., Lee J., Yoo A., "Predicting the three-dimensional structures of proteins: combined alignment approach", *J. Kor. Phys. Soc.* **44**:611–616, 2004.

[34] Burgess A. W., Ponnuswamy P. K., Sheraga H. A., "Analysis of conformations of amino acid residues and prediction of backbone topography in proteins", *Israel J. Chem.* **12**:239–286, 1974.

[35] Chou P. Y., Fasman G. D., "Prediction of protein conformation", *Biochemistry* **13**:222–245, 1974.

[36] Chou P. Y., Fasman G. D., "Prediction of the secondary structure of proteins from their amino acid sequence", *Adv. Enzymol. Relat. Areas Mol. Biol.* **47**:45–147, 1978.

[37] Lim V. I., "Algorithms for prediction of alpha-helical and beta-structural regions in globular proteins", *J. Mol. Biol.* **88**:873–894, 1974.

[38] Kabsch W., Sander C., "How good are predictions of protein secondary structure?", *FEBS Lett.* **155**:179–182, 1983.

[39] Kyngas J., Valjakka J., "Unreliability of the Chou-Fasman parameters in predicting protein secondary structure", *Protein Engineering* **11**:345–348, 1998.

[40] Garnier J., Osguthorpe D. J., Robson B., "Analysis and implications of simple methods for predicting the secondary structure of globular proteins", *J. Mol. Biol.* **120**:97–120, 1978.

[41] Gibrat J. F., Garnier J., Robson B., "Further developments of protein secondary structure prediction using information theory – new parameters and consideration of residue pairs", *J. Mol. Biol.* **198**:425–443, 1987.

[42] Garnier J., Gibrat J. F., Robson B., "GOR method for predicting protein secondary structure from amino acid sequence", *Methods Enzymol.* **266**:540–553, 1996.

[43] Kloczkowski A., Ting K. L., Jernigan R. L., Garnier J., "Protein secondary structure prediction based on the GOR algorithm incorporating multiple sequence alignment information", *Polymer* **43**:441–449, 2002.

[44] Kloczkowski A., Ting K. L., Jernigan R. L., Garnier J., "Combining the GOR V algorithm with evolutionary information for protein secondary structure prediction from amino acid sequence", *Proteins: Struct. Funct. Genet.* **49**:1554–166, 2002.
URL http://gor.bb.iastate.edu

[45] Sadeghi M., Parto S., Arab S., Ranjbar B., "Prediction of protein secondary structure based on residue pair types and conformational states using dynamic programming algorithm", *FEBS Letters* **579**:3397–3400, 2005.

[46] Holley L. H. and Karplus M., "Protein secondary structure prediction with a neural network", *Proc. Natl. Acad. Sci.* **86**:152–156, 1989.

[47] Kneller D. G., Cohen F. E., Langridge R., "Improvements in protein secondary structure prediction by an enhanced neural network", *J. Mol. Biol.* **214**:171–182, 1990.

[48] Zhang X., Mesirov J. P. and Waltz D. L., "Hybrid system for protein secondary structure prediction", *J. Mol. Biol.* **225**:1049–1063, 1992.

[49] Rost B. Sander C., "Prediction of protein secondary structure at better than 70% accuracy", *J. Mol. Biol.* **232**:584–599, 1993.

[50] Barlow T. W., "Feed-forward neural networks for secondary structure prediction", *J. Mol. Graph.* **13**:175–183, 1995.

[51] Jones Davy T., "Proteins secondary structure prediction based on position-specific scoring matrices", *J. Mol. Biol.* **292**:195–202, 1999.
URL http://bioinf.cs.ucl.ac.uk/psipred

[52] Cuff J. A., Barton G. J., "Application of enhanced multiple sequence alignment profiles to improve protein secondary structure prediction", *Proteins: Struct. Funct. Genet.* **40**:502–511, 2000.

[53] Petersen T. N., Lundegaard C., Nielsen M., Bohr H., Bohr J., Brunak S., Gippert G. P., Lund O., "Prediction of Protein Secondary Structure at 80% Accuracy", *Proteins: Struct. Funct. Genet.* **41**:17–20, 2000.

[54] Lin K., Simossis V. A., Taylor W. R., Heringa J., "A simple and fast secondary structure prediction method using hidden neural networks", *Bioinformatics* **21**(2):152–159, 2005.
URL http://www.ibi.vu.nl/programs/yaspinwww

[55] Pollastri G., Przybylski D., Rost B., Baldi P., "Improving the prediction of secondary structure in three and eight classes using recurrent neural networks and profiles", *Proteins: Struct. Funct. Genet.* **47**:228–235, 2002.
URL http://scratch.proteomics.ics.uci.edu

[56] Pollastri G., McLysaght A., "Porter: a new, accurate server for protein secondary structure prediction", *Bioinformatics* **21**(8):1719–1720, 2005.
URL http://distill.ucd.ie/porter

[57] Pollastri G., Martin A. J. M., Mooney C., Vullo A., "Accurate prediction of protein secondary structure and solvent accessibility by consensus combiners of sequence and structure information", *BMC Bioinformatics* **8**, 2007.

[58] Adamczak R., Porollo A., Meller J., "Combining Prediction of Secondary Structure and Solvent Accessibility in Proteins", *Proteins: structure, function and bioinformatics* **59**:467–475, 2005.
URL http://sable.cchmc.org

[59] Wood M. J., Hirst J. D., "Protein secondary structure prediction with dihedral angles", *Proteins: structure, function and bioinformatics* **59**:476–481, 2005.

[60] Zhang G. Z., Huang D. S., Zhu Y. P., Li Y. X., "Improving protein secondary structure prediction by using the residue conformational classes", *Pattern Recognition Letters* **26**:2346–2352, 2005.

[61] Chen J., Chaudhari N. S., "Bidirectional segmented-memory recurrent neural network for protein secondary structure prediction", *Soft Computing* **10**:315–324, 2006.

[62] Dor O., Zhou Y. Q., "Achieving 80% ten-fold cross-validated accuracy for secondary structure prediction by large-scale training", *Proteins: structure, function and bioinformatics* **66**:838–845, 2007.

[63] Zvelebil M. J. J. M., Barton G. J., Taylor W. R., Sternberg M. J. E., "Prediction of protein secondary structure and active sites using the alignment of homologue sequences", *J. Mol Biol.* **195**:957–961, 1987.

[64] Yi T. and Lander E. S., "Protein secondary structure prediction using nearest-neighbor methods", *J. Mol. Biol.* **232**:1117–1129, 1993.

[65] Geourjon C., Deleage G., "SOPM: a self-optimized method for protein secondary structure prediction", *Protein Engineering* **7**(2):157–164, 1994.

[66] Geourjon C., Deleage G., "SOPMA: significant improvements in protein secondary structure prediction by consensus prediction from multiple alignments", *Computer Applications in the Biosciences* **11**(6):681–684, 1995.

[67] Salamov A. A., Solovyev V. V., "Prediction of protein secondary structure by combining nearest-neighbor algorithms and multiple sequence alignments", *J. Mol. Biol.* **247**:11–15, 1995.

[68] Goldman N., Thorne J. L., Jones D. T., "Using evolutionary trees in protein secondary structure prediction and other comparative sequence analyses", *J. Mol. Biol.* **263**:196–208, 1996.

[69] Levin J. M., "Exploring the limits of nearest neighbour secondary structure prediction", *Protein Engineering* **10**:771–776, 1997.

[70] Salamov A. A., Solovyev V. V., "Protein secondary structure prediction using local alignments", *J. Mol. Biol.* **268**:31–36, 1997.

[71] Rychlewski L. Godzik A., "Secondary structure prediction using segment similarity", *Protein Engineering* **10**:1143–1153, 1997.

[72] Frishman D., Argos P., "Incorporation of non-local interactions in protein secondary structure prediction from the amino acid sequence", *Protein Engineering* **9**:133–142, 1996.

[73] Frishman D., Argos P., "Seventy-five percent accuracy in protein secondary structure prediction", *Proteins: Struct. Funct. Genet.* **27**:329–335, 1997.

[74] Aydin Z., Altunbasak Y., Borodovsky M., "Protein secondary structure prediction for a single-sequence using hidden semi-Markov models", *BMC Bioinformatics* **7**, 2006.
URL http://exon.biology.gatech.edu/ipssp/webIPSSP.cgi

[75] Martin J., Gibrat J. F., Rodolphe F., "Analysis of an optimal hidden Markov model for secondary structure prediction", *BMC Struct. Biol.* **6**, 2006.

[76] Won K. J., Hamelryck T., Prügel-Bennett A., Krogh A., "An evolutionary method for learning HMM structure: prediction of protein secondary structure", *BMC Bioinformatics* **8**, 2007.

[77] Yao X. Q., Zhu H. Q., She Z. S., "A dynamic Bayesian network approach to protein secondary structure prediction", *BMC Bioinformatics* **9**, 2008.

[78] Abe N., Mamitsuka H., "Predicting protein secondary structure using stochastic tree grammars", *Machine Learning* **29**:275–301, 1997.

[79] Hua S., Sun Z., "A novel method of protein secondary structure prediction with high segment overlap measure: support vector machine approach", *J. Mol Biol.* **308**:397–407, 2001.

[80] Kim H., Park H., "Protein secondary structure prediction based on an improved support vector machines approach", *Protein Engineering* **16**:553–560, 2003.

[81] Ward J. J., McGuffin L. J., Buxton B. F., Jones D. T., "Secondary structure prediction with support vector machines", *Bioinformatics* **19**:1650–1655, 2003.

[82] Guo J., Chen H., Sun Z., Lin Y., "A novel method for protein secondary structure prediction using dual-layer SVM and profiles", *Proteins: structure, function and bioinformatics* **54**:738–743, 2004.
URL http://www.bioinfo.tsinghua.edu.cn/pmsvm

[83] Liu Y., Carbonell J., Klein-Seetharaman J., Gopalakrishnan V., "Comparison of probabilistic combination methods for protein secondary structure prediction", *Bioinformatics* **20**:3099–3107, 2004.

[84] Birzele F., Kramer S., "A new representation for protein secondary structure prediction based on frequent patterns", *Bioinformatics* **22**:2628–2634, 2006.

[85] Pan X., "Multiple linear regression for protein secondary structure prediction", *Proteins: Struct. Funct. Genet.* **43**(3):256–259, 2001.

[86] Armano G., Mancosu G., Milanesi L., Orro A., Saba M., Vargiu E., "A hybrid genetic-neural system for predicting protein secondary structure", *BMC Bioinformatics* **6**, 2005.

[87] Cheng J., Randall A., Sweredoski M., Baldi P., "SCRATCH: a protein structure and structural feature prediction server", *Nucleic Acids Research* web server issue **33**:72–76, 2005.

[88] Bondugula R., Xu D., "MUPRED: a tool for bridging the gap between template based methods and sequence profile based methods for protein secondary structure prediction", *Proteins: structure, function and bioinformatics* **66**:664–670, 2007.
URL http://digbio.missouri.edu/mupred/ssp_server.html

[89] Guermeur Y., Geourjon C., Gallinari P., Deleage G., "Improved performance in protein secondary structure prediction by inhomogeneous score combination", *Bioinformatics* **15**:413–421, 1999.

[90] Ouali M., King R. D., "Cascaded multiple classifiers for secondary structure prediction", *Protein Science* **9**:1162–1176, 2000.

[91] Albrecht M., Tosatto S. C. E., Lengauer T., Valle G., "Simple consensus procedures are effective and sufficient in secondary structure prediction", *Protein Engineering* **16**:459–462, 2003.

[92] Guermeur Y., Pollastri G., Elisseeff A., Zelus D., Paugam-Moisy H., Baldi P., "Combining protein secondary structure prediction models with ensemble methods of optimal complexity", *Neurocomputing* **56**:305–327, 2004.

[93] Sen T. Z., Cheng H. T., Kloczkowski A., Jernigan R. L., "A consensus data mining secondary structure prediction by combining GOR V and fragment database mining", *Protein Science* **15**:2499–2506, 2006.
URL http://gor.bb.iastate.edu/cdm

[94] Cole C, Barber J. D. and Barton G. J., "The Jpred 3 secondary structure prediction server", *Nucleic Acids Research* **36**, 2008.
URL http://www.compbio.dundee.ac.uk/~www-jpred

[95] Eyrich V. A., Martí-Renom M. A., Przybylski D., Madhusudhan M. S., Fiser A., Pazos F., Valencia A., Sali A. and Rost B., "EVA: continuous automatic evaluation of protein structure prediction servers", *Bioinformatics* **17**(12):1242–1243, 2001.

[96] Rost B., Eyrich V. A., "EVA: large-scale analysis of secondary structure prediction", *Proteins* suppl. **5**:192–199, 2001.
URL http://cubic.bioc.columbia.edu/eva/sec/res_sec.html

[97] Koh I. Y., Eyrich V. A., Martí-Renom M. A., Przybylski D., Madhusudhan M. S., Eswar N., Graña O., Pazos F., Valencia A., Sali A., Rost B., "EVA: evaluation of protein structure prediction servers", *Nucleic Acids Research* **31**(13):3311–3315, 2003.

[98] Berman H. M., Westbrook J., Feng Z., Gilliland G., Bhat T. N., Weissig H., Shindyalov I. N., Bourne P.E., "The Protein Data Bank", *Nucleic Acids Research* **28**:235–242, 2000.

[99] Raghava G. P. S., "APSSP2: a combination method for protein secondary structure prediction based on neural network and example based learning", *CASP5* pages A–132, 2002.
URL http://www.imtech.res.in/raghava/apssp2

[100] Rost B., "PHD: predicting one-dimensional protein structure by profile based neural networks", *Methods Enzymol.* **266**:525–539, 1996.
URL http://cubic.bioc.columbia.edu/predictprotein

[101] Xu Y., Xu D., "Protein threading using PROSPECT: design and evaluation", *Proteins: Struct. Funct. Genet.* **40**:343–354, 2000.
URL http://compbio.ornl.gov/structure/prospect2

[102] Karplus K., Barrett C. and Hughey R., "Hidden Markov Models for detecting remote protein homologies", *Bioinformatics* **14**:846–856, 1998.
URL http://compbio.soe.ucsc.edu/SAM_T08/T08-query.html

[103] Chu W., Ghahramani Z., Podtelezhnikov A. and Wild D. L., "Bayesian segmental models with multiple sequence alignment profiles for protein secondary structure and contact map prediction", *IEEE/ACM Trans. Comput. Biol. Bioinformatics* **3**(2):98–113, 2006, ISSN 1545-5963, doi: http://dx.doi.org/10.1109/TCBB.2006.17.
URL http://wsbc.warwick.ac.uk/www/eva/submiteva.html

[104] Cuff J. A., Barton G. J., "Evaluation and improvement of multiple sequence methods for protein secondary structure prediction", *Proteins: Struct. Funct. Genet.* **34**:508–519, 1999.

[105] Meiler J., Mueller M., Zeidler A., Schmaeschke F., "Generation and evaluation of dimension-reduced amino acid parameter representations by artificial neural networks", *J. Mol. Model.* **7**:360–369, 2001.
URL http://www.meilerlab.org/web/view.php

[106] Rohl C. A., Strauss C. E., Misura K. M., Baker D., "Protein structure prediction using Rosetta", *Methods Enzymol.* **383**:66–93, 2004.

[107] Keller J., Gray M. and Givens J., "A fuzzy k-nearest neighbor algorithm", *IEEE Trans. Syst. Man and Cybernet.* **15**:580–585, 1985.

[108] Vapnik V. N., *Statistical Learning Theory*, Wiley, 1998.

[109] Sixma T. K., Van Zanten B. A. M., Dauter Z., Hol W. G. J., "Refined structure of Escherichia coli heat-labile enterotoxin, a close relative of cholera toxin", *J. Mol. Biol.* **230**:890–918, 1993.

[110] Chandonia J. M., Walker N. S., Lo Conte L., Koehl P., Levitt M., Brenner S.E., "ASTRAL compendium enhancements", *Nucleic Acids Research* **30**:260–263, 2002.

[111] Brenner S. E., Koehl P., Levitt M., "The ASTRAL compendium for sequence and structure analysis", *Nucleic Acids Research* **28**:254–256, 2000.

[112] Kabsch W., Sander C., "A dictionary of protein secondary structure: pattern recognition of hydrogen-bonded and geometrical features", *Biopolymers* **22**:2577–2637, 1983.

[113] Frishman D., Argos P., "Knowledge-based protein secondary structure assignment", *Proteins* **23**:566–579, 1995.

[114] Richards F. M. and Kundrot C. E., "Identification of structural motifs from protein coordinate data: secondary structure and first-level supersecondary structure", *Proteins* **3**:71–84, 1988.

[115] Kihara D., "The effect of long-range interactions on the secondary structure formation of proteins", *Protein Science* **14**:1955–1963, 2005.

[116] Gromiha M. M., Selvaraj S., "Influence of medium and long range interactions in protein folding", *Prep. Biochem. and Biotech.* **29**:339–351, 1999.

[117] Riedesel H., Kolbeck B., Schmetzer O., Knapp E. W., "Peptide binding at class I major histocompatibility complex scored with linear functions and support vector machines", *Genome Informatics* **15**:198–212, 2004.

[118] Fisher R. A., "The use of multiple measurements in taxonomic problems", *Annals of Eugenics* **7**:179–188, 1936.

[119] Rost B., Sander C., Schneider R., "Redefining the goals of protein secondary structure prediction", *J. Mol. Biol.* **235**(1):13–26, 1994.

[120] Gorodkin J., "Comparing two K-category assignments by a K-category correlation coefficient", *Comp. Biol. and Chem.* **28**:367–374, 2004.

[121] Rost B., "Rising accuracy of protein secondary structure prediction", *Protein structure determination, analysis, and modeling for drug discovery* pages 207–249, 2003.

[122] Henikoff S., "Amino acid substitution matrices from protein blocks", *Proc. Nat. Acad. Sci.* **89**:10915–10919, 1992.

[123] Dayhoff M. O., Schwartz R. and Orcutt B. C., "A model of evolutionary change in proteins", *Atlas of protein sequence and structure* **5**:345–358, 1978.

[124] Gribskov M., McLachlan A. D., Eisenberg D., "Profile analysis: detection of distantly related proteins", *Proc. Natl. Acad. Sci.* **84**(13):4355–4358, 1987.

[125] Gribskov M., Luthy R., Eisenberg D., "Profile analysis", *Methods Enzymol.* **183**:146–159, 1990.

[126] Gribskov M., Veretnik S., "Identification of sequence pattern with profile analysis", *Methods Enzymol.* **266**:198–212, 1996.

[127] Niermann T., Kirschner K., Crawford I. P., "Prediction of tertiary structure of the alpha-subunit of tryptophan synthase", *Biol. Chem.* **368**:1087–1088, 1987.

[128] Benner S. A., Gerloff D., "Patterns of divergence in homologous proteins as indicators of secondary and tertiary structure: a prediction of the structure of the catalytic domain of protein kinases", *Adv. Enzyme Reg.* **31**:121–181, 1990.

[129] Press W. H., Flannery B. P., Teukolsky S. A., W. T. Vetterling, *Numerical Recipes: The Art of Scientific Computing*, Cambridge University Press, 1986.

[130] Carl D. Meyer, *Matrix Analysis and Applied Linear Algebra*, SIAM, 2000.

[131] Goldberg M. A., *An Introduction to Probability Theory*, Plenum Press, 1984.

[132] Pearson K., "On Lines and Planes of Closest Fit to Systems of Points in Space", *Philosophical Magazine* **2**(6):559–572, 1901.

[133] Comon P., "Independent Component Analysis: a new concept?", *Signal Processing* **36**(3):287–314, 1994.

[134] Altschul S. F., Madden T. L., Schaffer A. A., Zhang J., Zhang Z., Miller W., and Lipman D. J., "Gapped BLAST and PSI-BLAST: a new generation of protein database search programs", *Nucleic Acids Research* **25**(17):3389–3402, 1997.

[135] Benson D. A., Karsch-Mizrachi I., Lipman D. J., Ostell J., Wheeler D. L., "GenBank", *Nucleic Acids Research* **36**(Database issue):D25–30, 2008.

[136] Bairoch A., Boeckmann B., Ferro S., Gasteiger E., "Swiss-Prot: juggling between evolution and stability", *Brief. Bioinform.* **5**:39–55, 2004.

[137] Wu C. H., Yeh L. S., Huang H., Arminski L., Castro-Alvear J., Chen Y., Hu Z. Z., Ledley R. S., Kourtesis P., Suzek B. E., Vinayaka C. R., Zhang J., Barker W.C., "The Protein Information Resource", *Nucleic Acids Research* **31**:345–347, 2003.

[138] Aimoto S., Ono S., editor, *Peptide Science 2007*, 2008.

[139] Pruitt K. D., Tatusova, T., Maglott D. R., "NCBI Reference Sequence (RefSeq): a curated non-redundant sequence database of genomes, transcripts and proteins", *Nucleic Acids Research* **35**(Database issue):D61–65, 2007.

[140] Simon Haykin, *Neural Networks*, Prentice Hall, 1999.

[141] Altschul S. F., Gish W., Miller W., Myers E. W., Lipman D. J., "Basic local alignment search tool", *J. Mol. Biol.* **215**(3):403–410, 1990.

[142] Przybylski D., Rost B., "Alignments grow, secondary structure prediction improves", *Proteins: Struct. Funct. Genet.* **46**:197–205, 2002.
URL http://cubic.bioc.columbia.edu/predictprotein

i want morebooks!

Buy your books fast and straightforward online - at one of world's fastest growing online book stores! Environmentally sound due to Print-on-Demand technologies.

Buy your books online at
www.get-morebooks.com

Kaufen Sie Ihre Bücher schnell und unkompliziert online – auf einer der am schnellsten wachsenden Buchhandelsplattformen weltweit! Dank Print-On-Demand umwelt- und ressourcenschonend produziert.

Bücher schneller online kaufen
www.morebooks.de

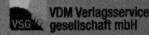

VDM Verlagsservicegesellschaft mbH
Heinrich-Böcking-Str. 6-8   Telefon: +49 681 3720 174   info@vdm-vsg.de
D - 66121 Saarbrücken   Telefax: +49 681 3720 1749   www.vdm-vsg.de

Printed by Books on Demand GmbH, Norderstedt / Germany